D0759619

Molluscan
Nerve Cells:
FROM BIOPHYSICS
TO BEHAVIOR

REPORT OF A MEETING SUPPORTED BY
The Marie H. Robertson Fund for Neurobiology

Cold Spring Harbor Reports in the Neurosciences • Volume 1

Molluscan Nerve Cells:
FROM BIOPHYSICS TO BEHAVIOR

Edited by

JOHN KOESTER

College of Physicians & Surgeons
of Columbia University

JOHN H. BYRNE

University of Pittsburgh
School of Medicine

Cold Spring Harbor Laboratory
1980

Cold Spring Harbor Reports in the Neurosciences

Volume 1 • Molluscan Nerve Cells: From Biophysics to Behavior

Molluscan Nerve Cells: From Biophysics to Behavior
© 1980 by Cold Spring Harbor Laboratory
Printed in the United States of America
Cover and book design by Emily Harste

Library of Congress Cataloging in Publication Data

Main entry under title:

Molluscan nerve cells, from biophysics to behavior.

  (Cold Spring Harbor reports in the neurosciences
; v. 1)
  Report of a meeting supported by the Marie H.
Robertson Fund for Neurobiology, held at Cold Spring
Harbor Laboratory, May 18-21, 1980.
  Includes index.
    1. Action potentials (Electrophysiology)—Con-
gresses. 2. Neural transmission—Congresses.
3. Mollusks—Cytology—Congresses. 4. Nervous
system—Mollusks—Congresses. 5. Neurobiology.
I. Koester, John. II. Byrne, John H. III. Series.
[DNLM: 1. Mollusca—Cytology—Congresses.
2. Neurons—Congresses. W1 C0133E v. 1 / QX 675
M727 1980]
QP363.M62   594'.0188   80-39967
ISBN 0-87969-135-2

Participants

Floyd J. Brinley
Neurological Disorders Program
National Institute of Neurological Communication Disorders and Stroke
Bethesda, Maryland 20205

Arthur M. Brown
Department of Physiology
University of Texas Medical Branch
Galveston, Texas 77550

John H. Byrne
Department of Physiology
University of Pittsburgh School of Medicine
Pittsburgh, Pennsylvania 15261

John A. Connor
Department of Physiology and Biophysics
University of Illinois
Urbana, Illinois 61801

Susumu Hagiwara
Department of Physiology
UCLA School of Medicine
Los Angeles, California 90024

Eric Kandel
Division of Neurobiology and Behavior
Department of Physiology
College of Physicians & Surgeons of Columbia University
New York, New York 10032

Marc Klein
Division of Neurobiology and Behavior
Department of Physiology
College of Physicians & Surgeons of Columbia University
New York, New York 10032

John Koester
Division of Neurobiology and Behavior
Department of Physiology
College of Physicians & Surgeons of Columbia University
New York, New York 10032

Rudolfo Llinás
Department of Physiology & Biophysics
New York University Medical Center
New York, New York 10016

H. Dieter Lux
Max-Planck-Institut für Psychiatrie
Munich, Federal Republic of Germany 401240

Alain Marty
Laboratoire de Neurobiologie
Ecole Normale Supérieure
Paris, France 75230

Robert W. Meech
Department of Physiology
University of Utah
Salt Lake City, Utah 84112

Eli Shapiro
Division of Neurobiology and Behavior
Department of Physiology
College of Physicians & Surgeons of Columbia University
New York, New York 10032

Stephen Smith
Department of Physiology and Anatomy
University of California
Berkeley, California 94720

Thomas J. Smith, Jr.
Laboratory of Neurophysiology
National Institute of Neurological and Communicative Disorders and Stroke
Bethesda, Maryland 20205

Charles F. Stevens
Department of Physiology
Yale University School of Medicine
New Haven, Connecticut 06510

Roger C. Thomas
Department of Physiology
Yale University School of Medicine
New Haven, Connecticut 06510

Stuart Thompson
Department of Biology and Hopkins Marine Station
Stanford University
Stanford, California 94305

Douglas Tillotson
Department of Physiology
Boston University Medical School
Boston, Massachusetts 02118

Wilkie A. Wilson
Epilepsy Center
Veteran's Administration Hospital
Durham, North Carolina 27705

Transcribing Editors

David Harris
Division of Neurobiology and Behavior
Department of Physiology
College of Physicians & Surgeons of Columbia University
New York, New York 10032

Steven Mackey
Division of Neurobiology and Behavior
Department of Physiology
College of Physicians & Surgeons of Columbia University
New York, New York 10032

Judith Strong
Department of Physiology
Yale University School of Medicine
New Haven, Connecticut 06510

E. Terrell Walters
Division of Neurobiology and Behavior
Department of Physiology
College of Physicians & Surgeons of Columbia University
New York, New York 10032

Gary Yellen
Department of Physiology
Yale University School of Medicine
New Haven, Connecticut 06510

First row: J.H. Byrne, N. Dumser, J. Koester; Session in Banbury Center Meeting House
Second row: R. Llinás, E. Kandel; F.J. Brinley, E.T. Walters, D. Tillotson
Third row: A. Marty, H.D. Lux; M. Klein, C.F. Stevens
Fourth row: Participants at 1980 Molluscan Nerve Cell Meeting

Foreword

Over the past decade the Cold Spring Harbor Laboratory has steadily increased its programs in the neurosciences, initially through its holding of advanced summer courses and now also through the sponsoring of specific meetings and the carrying out of year-round research. These efforts reflect a resurgence of a fundamental interest expressed at two earlier phases in our history. During the late 1890s a magnificent new laboratory to study the invertebrates living in our nearby waters was built by our enlightened founder, Mr. John Devine Jones, a leader in the development of the American insurance industry. Among the early research workers in the Jones Laboratory (1899) was Gertrude Stein, then a medical student at John Hopkins who had studied zoology at Radcliffe College. Her mentor here was Charles Davenport, a product of the post–Louis Agassiz era at Harvard, who in 1904 was to move here permanently from the University of Chicago to assume the directorship of the Carnegie Institution of Washington's newly established unit at Cold Spring Harbor. Then, Mendel's Laws had just been rediscovered (1901), and the unexpectedly large resources that had just been put at Davenport's disposal from the steel mill wealth of Andrew Carnegie were used for a long-term commitment to a further working out of the fundamental rules of heredity.

Davenport's broad intellectual interests never were restricted to genetics, and in the early 1920s he worked hard to persuade the National Research Council to use funds coming from the Rockefeller Foundation to set up at Cold Spring Harbor a major new laboratory to study the biophysical properties of cells. Though he failed in this endeavor and the building went instead to Woods Hole, he was able subsequently to attract the generosity of many of the wealthy estate owners around Cold Spring Harbor. They founded the Long Island Biological Association in 1924 and appointed Reginald Harris as Director of its Biological Laboratories, as distinct from the unit financed by the Carnegie Institution. Harris quickly presided over the renovation of several buildings used in the summer teaching program of the Biological Laboratories, as well as over the erection of three new laboratories that became Davenport, Nichols, and James. In them, cell physiology, as opposed to genetics, was to be emphasized. Particularly important was the creation in 1929 of the Biophysics Section under the supervision of the Danish-born physicist Hugo Fricke, whose past research at the Cleveland Clinic had established, through measurements of the electrical capacitance, the ~50 Å thickness of the red blood cell membrane. Howard Curtis joined Fricke's staff in 1932, and during several subsequent summers Kenneth Cole, then at P&S, began a collaboration at Cold Spring Harbor with Curtis that eventually led to their seminal ideas on the propagation of electrical impulses along the axons of nerve cells.

Equally important was the idea of Reginald Harris to bring together for lengthy periods each summer a number of distinguished scientists to discuss topics on "quantitative" as opposed to descriptive aspects of biology. The first such Cold Spring Harbor "Symposium," backed by support from the Rockefeller Foundation, was held in 1933 and brought together some 30 physicists, chemists, and biologists to talk about "Surface Behavior." In 1934, "Aspects of Growth" was the topic, with "Photochemical Reactions" being the focus of attention in 1935, to be followed in 1936 by "Excitation Phenomena."

By then, the continuing economic depression was reflected in an almost total loss of financial support from the local estate families, and with Harris's unexpected early death from pneumonia in the winter of 1936, the scientific staff of the Biological Laboratories was effectively reduced to Fricke and Eric Ponder, a Scottish-born cell physiologist whose research centered on the permeability of the red blood cells. Ponder functioned briefly as Director of the Biological Laboratories until 1940, attracting as summer visitors several later-to-be-prominent biologists, including James Danielli, Hugh Davson, Hans Neurath, and Walter Rosenblith. Then the steadily deteriorating financial structure of the Biological Laboratories led necessarily to its effective amalgamation with the Department of Genetics of the Carnegie Institution. Year-round research at Cold Spring Harbor returned to an exclusive, and soon to be appreciated, totally successful preoccupation with genetics under the direction of Miloslav Demerec.

Neurophysiological phenomena, however, continued to be the major concern of several of the annual Symposia. The 1952 gathering focused on "The Neuron" and provided the first occasion for a large-scale discussion of the newly formulated Hodgkin-Huxley model of nerve conduction. In 1965 "Sensory Receptors" were emphasized, reflecting Max Delbruck's deep interest in sensory transducers. Several years later, in 1968, Delbruck organized a summer workshop on "Sensory Transducers" where bacterial chemotaxis, the responses of the mould *Phycomyces* to light and the male silkworm to the female sex lure, and the mechanosensitivity of the motoric cilia on the gills of the mussel *Mytilus* were examined.

Soon afterward, a grant proposal was submitted to the Sloan Foundation for funds to renovate much of the by then badly run down "Animal House" into well-equipped laboratories for the teaching of advanced summer courses in the neurosciences. This proposal, which also requested monies to support the teaching efforts themselves, was made knowing that the Sloan Foundation was embarking upon a major thrust in neuroscience. Happily, our proposal was judged favorably, and in 1970 the "Animal House," soon to be renamed McClintock Laboratory, was radically modernized, allowing us to offer in 1971 both a course on Electrophysiological Methods for Cellular Neurobiology that focused on *Aplysia*, as well as a Neurobiology lecture course aimed at students coming from other disciplines. Upon expiration of our Sloan Funds, the receipt in 1974 of a Federal Training grant allowed continuation of the program, which by then had grown to three courses each summer. By 1976, however, uncertainties about whether training funds should be used to support such summer efforts began to give us many anxious moments.

That our program not only continued but actually grew in scope was possible only because of support given us by our neighbor Mr. Charles S. Robertson and his family. Through the formation of the well-endowed Robertson Research Fund, monies became available for (1) the creation of new research facilities, either through renovation or new construction; (2) the purchase of major new items of equipment; and (3) the awarding of fellowships and summer course stipends. Using Robertson funds, we totally renovated in 1975 the J.D. Jones Laboratory for year-round use by neurobiologists. The new facility was first used during the summer of 1976 for an experimental course on the CNS, as well as for a CNS workshop supported by Sloan Foundation monies given specifically for three years of summer workshops. In 1978 a summer course in Jones on Neuroanatomical Methods commenced, and in 1979 an advanced workshop devoted to the neurobiological aspects of pain came into existence with the help of the Rita Allen Foundation.

Even further expansion was made possible when, in 1976, Mr. Robertson deeded to the Laboratory the major components of his estate, including his home, Robertson House, located some three miles from us in Lloyd Harbor. By creating there the Banbury Conference Center, we now have facilities for the holding of small lecture courses and meetings that are virtually unmatched anywhere. Over the past several summers, we have held there our lecture courses on the Synapse, the Neurobiology of Behavior, and the Principles of Neural Development, as well as several Sloan Foundation sponsored workshops in the cognitive sciences.

In addition, through the Banbury Foundation, the Robertson family has created the Marie H. Robertson Fund for Neurobiology in memory of Mr. Robertson's wife, who died in 1972. Established in 1976, this fund at first was largely used to maintain our summer teaching program in the face of what was at best erratic federal funding. With the receipt from NIH in 1979 of a substantial training grant, and with additional support from NSF, the decision was made to use some of the Marie H. Robertson funds to help support specific summer workshops as well as to fund one to two specialized meetings each year at the Banbury Conference Center.

Helping us draw up specific plans for these workshops and meetings have been the members of our Neurobiology Advisory Committee, David Hubel, Eric Kandel, John Nichols, and Charles Stevens. We remain much in their debt for their counsel, which included the proposal to focus our first Robertson-supported meeting on Molluscan Nerve Cells. Its organizers, Charles Stevens and Eric Kandel, are to be congratulated for the excellence with which this proposal was brought to fruition. We are also much indebted to John Koester and John Byrne for their tireless efforts as editors of the report of the meeting, the first of what we hope will be a distinguished series of *Reports in the Neurosciences*.

I must also acknowledge our Administrator Director, William Udry, for his most effective efforts over the past decade in finding ways to keep open the possibility of federal funds being committed to our program. We are also indebted to Birgit Zipser, the first neurobiologist to serve on our staff during this era, who of necessity has had to bear solely all too many of the administrative chores that go with a program of our current size.

J. D. Watson

Preface

Two areas of neurobiology in which there has been remarkable progress during the last decade are membrane biophysics and cellular analysis of behavior. To a large degree, the recent advances in these two fields have resulted from the use of simple invertebrates, particularly annelids, arthropods, and molluscs. The experimental advantages of molluscan preparations in this regard are twofold. First, many molluscan species have giant neurons, with cell bodies and axons large enough to be impaled easily with microelectrodes for recording, stimulating, and voltage-clamping. Thus, these cells are ideal for studies of membrane phenomena that underlie neuronal signaling. Most molluscan species share a second experimental advantage—that of a relatively simple nervous system. The entire nervous system in some gastropods contains as few as 10,000 neurons. A single ganglion, capable of mediating a variety of behaviors, may be made up of only one thousand nerve cells. Therefore, it is often possible to relate individual identified cells to specific behaviors and changes in the activity of these cells to specific behavioral modifications.

Beginning over 30 years ago with the classic studies of Hodgkin, Huxley, and Katz, molluscan preparations were utilized originally to examine the ionic mechanisms underlying the resting potential and action potential, various aspects of synaptic transmission, and the regulation of intracellular ions. As a result of this earlier work and the continued progress in this area, a number of important phenomena have been identified that appear to be widespread throughout the animal kingdom. More recently, these preparations have also been utilized to examine the cellular basis of behavior. Here again, a number of principles have begun to emerge that are likely to prove generally applicable. Within the last several years, these two levels of analysis have begun to converge, so that it is now possible to relate the biophysical properties of individual neurons to the features of the behavior that they mediate.

To stimulate interaction between investigators in these two areas of molluscan neurobiology, Eric Kandel and Charles Stevens organized a 3-day meeting that was held at Cold Spring Harbor Laboratory in May 1980. This meeting brought together the discoverers of the newly described channels, as well as other major contributors who work on the biophysical properties of nerve cells, to assess recent progress in this field. This volume represents a summary of the talks given by the 20 participants. Only a few speakers submitted manuscripts of their presentations. Most were recorded and subsequently transcribed by a group of graduate-student scribes who also attended the conference. The result is a fairly broad survey of current work on the biophysics of molluscan nerve cells, including two examples of studies in which the control of behavior has been correlated with the membrane properties of individual, identified nerve cells.

This volume should prove useful for advanced neurobiological investigators. In addition, we have taken several steps to ensure that the material covered here also will be accessible to graduate students and to others who are just beginning to turn their interest to the study of membrane biophysics and behavior. Eric Kandel has written an introductory chapter that provides a brief review of the history of discovery of the various membrane phenomena described in later chapters. A brief description of the different experimental techniques used in the studies described in this book and in related work is given in Charles Stevens' chapter. In addition, many technical and quantitative details have been deleted from the transcriptions of the talks. This information can be obtained from the references at the end of each chapter.

The publication of this volume is due to the cooperative effort of several individuals. On behalf of the other participants, we thank James Watson for initiating, and Eric Kandel and Charles Stevens for organizing the meeting on which this book is based. We would also like to thank David Harris, Steven Mackey, Judith Strong, E. Terrell Walters, and Gary Yellen for their excellent job of transcribing the oral presentations. Finally, we are deeply indebted to Judy Cuddihy, Nadine Dumser, and Nancy Ford of the Publications Department at Cold Spring Harbor Laboratory for their first-rate editorial assistance and for their efforts to speed the publication of this book.

J. **Koester**
J. **Byrne**

Contents

Abbreviations

ELECTROPHYSIOLOGICAL TERMS

The symbols and subscripts in this section are used in various combinations throughout this volume. The subscripts are used to modify the symbols, as in the following examples: I_t, tail current; g_m, membrane conductance; V_h, holding potential.

Primary Symbols and Abbreviations

C, capacitance
E, equilibrium potential
emf, electromotive force
EPSP, excitatory postsynaptic potential
g, conductance
I, current
IPSC, inhibitory postsynaptic current
IPSP, inhibitory postsynaptic potential
IR, interposed repolarization
P, permeability
PSC, postsynaptic current
PSP, postsynaptic potential
R, resistance
τ, time constant
V, potential difference
$[\cdot]$, ionic concentration

Subscripts

A, as in I_A, A current, or fast K^+ current
Ca, K, or Na, calcium, potassium, or sodium ions
e, external
h, holding
i, internal
K(Ca), as in $I_{K(Ca)}$, Ca^{++}-activated I_K
K(V), as in $I_{K(V)}$, voltage-dependent or delayed rectification; I_K late
L, leakage
m, membrane
R, resting
rev, reversal
∞, steady state
t, tail
T, test

UNITS OF MEASURE

Prefixes

$m = 10^{-3} = milli\text{-}$
$\mu = 10^{-6} = micro\text{-}$
$n = 10^{-9} = nano\text{-}$
$p = 10^{-12} = pico\text{-}$

Units

$A = amps$
$F = farads$
$S = Siemens$
$V = volts$
$\text{Å} = 10^{-7}\ mm$

CHEMICAL COMPOUNDS

ACh, acetylcholine
3-AP, 3-aminopyridine
4-AP, 4-aminopyridine
ATPase, adenosine triphosphatase
cAMP, 3′, 5′ cyclic adenosine monophosphate
BDAC, bis(2-hydroxyethyl)dimethylammonium chloride
CCmP, carbonyl cyanide m-chlorophenyl hydrazone
CN, cyanide
EDTA, ethylene diaminetetraacetic acid
EGTA, ethyleneglycoltetraacetic acid
FCCP, carbonyl cyanide p-trifluoromethoxyphenylhydrazone
HEDTA, N′-(2-hydroxyethyl)-ethylenediamine-N, N, N′-triacetate
HEPES, N-2-hydroxyethylpiperazine-N′-2-ethane sulfonic acid
SITS, 4-acetamido-4-isothiocyanate-stilbene-2, 2′-disulfonic acid
Tris, tris(hydroxymethyl)aminomethane
TEA, tetraethylammonium
TTX, tetrodotoxin

Molluscan
Nerve Cells:
FROM BIOPHYSICS
TO BEHAVIOR

The Multichannel Model of the Nerve Cell Membrane: A Perspective

ERIC KANDEL

Division of Neurobiology and Behavior
Department of Physiology
College of Physicians & Surgeons
of Columbia University
New York, New York 10032

• Science does not advance smoothly and continuously but in fits and starts. Periods of creative productivity often are followed by periods of consolidation in which the general nature of the new findings is explored but relatively few novel ideas are generated. In neurobiology, alternating periods of excitement and calm can be seen in several areas, but they have been prominent particularly in the biophysical analysis of excitable membranes.

In the several years preceding and following the Second World War, the development of new conceptual and analytic tools—measurements of membrane impedance, intracellular recording of transmembrane potentials, and analysis of membrane currents by voltage clamp—led to a new view of the nerve cell membrane that rapidly took hold of our thinking and dominated it until recently. This new view—the dual channel model of excitable membranes—was derived from the analysis of the membrane of the axon of nerve cells. It proved so general that for a long while it seemed that membranes of the cell body, the terminals, and the dendrites must be made of the same material as the membrane of the axon. This tranquil confidence has been replaced by a new intellectual ardor in the last few years as we have been confronted rapidly with two sets of findings: (1) an entirely new family of ionic channels encountered in study of the cell body and terminal region and (2) the properties of elementary channel events.

I will try here to retrace some recent experiments that have led up to this new brilliant period of membrane biophysics, focussing primarily on the ionic currents that have been discovered in nerve

1

cells. In the next chapter, Charles Stevens describes methods used for studying ionic channels in nerve membranes. The presentations that follow illustrate the present state of these advances and how these advances have improved our understanding of nerve cell function, of transmitter release, and of behavior.

A TWO-CHANNEL MODEL ACCOUNTS FOR THE EXCITABILITY OF AXONS

• At the Paris Symposium on excitable membranes in 1949, Hodgkin, Huxley, and Katz startled the neurophysiological community with a simple, yet surprisingly coherent, quantitative model of the mechanisms of the action potential based on the initial analysis of experiments on the squid giant axon using voltage-clamp techniques (Hodgkin et al. 1949). According to this model, excitable properties of the axon could be understood in terms of only two independent channels, one for Na^+ and the other for K^+, both of which were gated by membrane voltage, being closed at rest and open with depolarization. With the appearance of the complete series of five papers in 1952 (Hodgkin and Huxley 1952a,b,c,d; Hodgkin et al. 1952), this model, expressed in the form of the ionic hypothesis, quickly became the central unifying principle of cellular neurobiology.

Earlier, Hodgkin and Katz (1949) had already outlined aspects of this theory. They confirmed that the resting potential (V_R) results from an unequal distribution of K^+, Na^+, and Cl^- across the membrane and assumed (as was later shown by Keynes and Hodgkin [1955]) that an energy-dependent Na^+-K^+ pump keeps the concentration of Na^+ inside the axon approximately ten times lower than that on the outside. The value of the membrane potential (V_m) in squid, typically −70 mV, is close to the Nernst equilibrium potential of K^+ (E_K). The small deviations from E_K were shown to be due to the slight leakage of the membrane to Na^+ and Cl^-. In addition, Hodgkin and Katz found that there is a sudden change in the permeability characteristics of the membrane during the action potential. Depolarization increases Na^+ permeability, which, in turn, causes further depolarization and in a regenerative process drives V_m toward the Na^+ equilibrium potential (E_{Na}) of about +55 mV.

By using the voltage-clamp technique developed by Cole (1949; Marmont 1949), which permitted changing V_m rapidly and maintaining it at any desired value, these ionic properties of the membrane could be better analyzed and quantified. The current that the feedback amplifier must supply to maintain V_m at a given value equals the net ionic current that flows at that V_m. By eliminating the Na^+ current (I_{Na}) selectively, Hodgkin and Huxley were able to separate the total current into specific ionic components carried by Na^+ and K^+. This analysis revealed that depolarization leads to the sequential opening first of Na^+ and then of K^+ channels, giving rise to an inward current followed by an outward current. The activation of the inward I_{Na} leads to a sudden reversal of V_m, but this reversal is transient and self-limiting. The progressive depolarization of the action potential ultimately shuts off (inactivates) the enhanced Na^+ permeability and leads to the increase in K^+ permeability. These two processes combine to restore V_m to the resting level.

In the 15 years that followed the original report of Hodgkin and Huxley, the two-channel model was shown to be remarkably general. It described well the excitability characteristics of a variety of axons in invertebrates, as well as the myelinated axons in vertebrates. It was natural to assume that the model also applied to other parts of the neuron. This assumption seemed to be confirmed by early voltage-clamp experiments of the cell body membrane. In two independent studies, one by Frank et al. (1959) and the other by Araki and Terzuolo (1962), it was found that depolarization of the cell body of the motor neuron in the cat spinal cord led to an inward current followed by an outward current. These currents had properties that seemed similar to the Na^+ (I_{Na}) and K^+ current (I_K) described in the squid. Frank and Tauc (1964) reached similar conclusions based on voltage-clamp studies of giant cell bodies of *Aplysia* neurons.

Thus, for a 15-year period, the excitable membranes of neurons seemed to be explained fully by a description of the properties and kinetics of the Na^+ channel, the K^+ channel, and the Na^+-K^+ pump. The pump was assumed to be electrically neutral, because it did not contribute to resting membrane potential (V_R) in the squid giant axon. It served exclusively to maintain ion gradients that provided the electrochemical potentials for the various ion species.

This view of the pump and of the channels in the cell body began to be questioned in the mid-1960s after two incidential discoveries: (1) that the pump was often electrogenic, and (2) that a new class of K^+ channels existed.

THE Na^+ PUMP CAN BE ELECTROGENIC

● Kerkut and Thomas (1964) demonstrated that in the cell body of snail neurons the Na^+-K^+ pump could generate an electrical potential. When the Na^+ concentration ($[Na^+]_i$) of the cell was raised artificially by intracellular injections, it produced a large hyperpolarization of V_m. This hyperpolarization was blocked by application of cardiac glycosides and by cooling procedures that block the Na^+-K^+ pump. Thomas later returned to this problem (Thomas 1972) and, using voltage-clamp techniques, measured directly the net current produced by the Na^+-K^+ pump, while simultaneously monitoring $[Na^+]_i$ with a Na^+-sensitive microelectrode. Thomas found that the pump did not contribute to V_m in the cells that he studied. Later, the pump was found to be active in other neurons, even with normal $[Na^+]_i$. In these cells, the pump contributes substantially to V_m. This was shown by Carpenter and Alving (1968) and by Gorman and Marmor (1970), who found that V_m can be dissociated experimentally into two components. One component arises from a membrane conductance, primarily to K^+, with properties similar to that found in squid giant axon. The other component depends upon an electrogenic Na^+ transport. This Na^+ component of the pump could contribute up to 30% of total V_R, or about $10-20$ mV.

THERE IS MORE THAN ONE CLASS OF K^+ CHANNELS

• In 1960, 4 years before Kerkut and Thomas discovered the electrogenicity of the Na^+ pump, Hagiwara et al. (1961) had pointed out an additional complexity in the membranes of neurons. When they voltage-clamped the cell bodies of neurons of the marine snail *Onchidium*, they found that in addition to the usual early inward current and delayed outward current (attributable to Na^+ and K^+, respectively), there was a second early outward current. This outward current had kinetics that were similar to those of the oppositely directed I_{Na}. This outward current was encountered later in other molluscan neurons and was analyzed in detail first by Connor and Stevens (1971a, b; Stevens 1969) and almost simultaneously by Neher and Lux (1971; Neher 1971). The current was shown to be a new type of I_K, which they called the early or fast K^+ current (or A current, I_A) to distinguish it from the conventional delayed I_K ($I_{K(V)}$), also called delayed rectification or the Hodgkin-Huxley K^+ current. The early I_K is activated rapidly and then inactivated when V_m is depolarized beyond -55 mV.

Discoveries of new membrane properties continued. Meech (1974) discovered a third class of K^+ channels that is activated by an increase of Ca^{++} on the inside (intracellular) surface of the membrane. It is now clear that this Ca^{++}-dependent I_K ($I_{K(Ca)}$) overlaps in time with the slow outward $I_{K(V)}$. Thus the late outward I_K is made up of two components, one a voltage-dependent component with the properties described by Hodgkin and Huxley, and the other a Ca^{++}-dependent component that is, at least in part, voltage-independent. This component is not turned on by depolarization alone, but by the influx of Ca^{++}. Thus, there is now evidence for at least three independent K^+ channels in the membrane of the cell body: $I_{K(V)}$, I_A, and $I_{K(Ca)}$.

INWARD CURRENT CAN ALSO BE CARRIED BY Ca^{++}

• In addition to several types of outward currents, another inward current was discovered in 1965 in the cell body of snail neurons by Gerasimov and his colleagues (Gerasimov et al. 1965). This inward current is carried by Ca^{++}. Geduldig and Junge (1968), working in Hagiwara's laboratory, found that the cell body of cell R2 in *Aplysia* had two inward currents, one carried by Na^+ and the other by Ca^{++}. The two currents were carried through two pharmacologically distinct classes of channels. I_{Na} was blocked selectively by tetrodotoxin (TTX), whereas the Ca^{++} component was selectively blocked by Co^{++}. Katz and Miledi (1969) have also presented evidence for two inward currents due to Na^+ and Ca^{++} in the synaptic terminal membrane at the squid giant synapse.

Not only are the activation properties for some of these new channels different from those described by Hodgkin and Huxley for I_{Na} and $I_{K(V)}$, but also their inactivation properties are different. Tillotson (1979) found that inactivation of the Ca^{++} channel is not dependent on voltage but on current: The inactivation results from the flow of Ca^{++} current (I_{ca}) through the channel.

NEW TYPES OF CHEMICALLY GATED CHANNELS HAVE ALSO BEEN DISCOVERED

• The channels we have considered so far are gated by voltage or, in the case of $I_{K(Ca)}$, by the influx of Ca^{++}. The membranes of nerve cells have, in addition, a variety of channels that are gated by chemical transmitters. Recently, several novel types of chemically gated channels have been encountered.

First, some channels have now been found that are open at the resting level of V_m and whose gates are closed by chemical transmitters. Initially described in studies of the sympathetic ganglion by Weight and Votava (1970), such decreased-conductance synaptic channels have now been demonstrated and analyzed in molluscan neurons by Gerschenfeld and Paupardin-Tritsch (1974a,b), Carew and Kandel (1976), and Byrne et al. (1979).

Second, until recently, voltage-gated channels and chemically gated channels were thought to be mutually exclusive categories. Voltage-gated channels were thought to be insensitive to transmitters and chemically gated channels were thought to be insensitive to membrane voltage. Recent work has shown that this distinction, although still useful, does not hold in all cases. For example, studies on nerve-muscle synapses of the frog by Anderson and Stevens (1973) and by others and on molluscan neurons by Wilson, Wachtel, and Pellmar (Pellmar and Wilson 1977; Wilson and Wachtel 1978) have shown that chemically sensitive channels are affected by voltage. Even more surprising is the finding that voltage-gated channels, such as those for I_{Ca} and $I_{K(V)}$, can be modulated strongly by chemical transmitters (Klein and Kandel 1978; Mudge et al. 1979; Shapiro et al. 1980).

Finally, there is now evidence that some slow, voltage-dependent synaptic actions seem to involve phosphorylation of channel proteins or associated regulatory components by cAMP-dependent protein kinases. These types of channels also occur in the presynaptic terminals where cyclic-nucleotide-mediated regulatory processes could serve as mechanisms for the short-term, and perhaps even long-term, control of transmitter release (Klein and Kandel 1978, 1980; Castellucci et al. 1980).

THE MEMBRANES OF THE CELL BODY AND OF THE PRESYNAPTIC TERMINAL ARE MORE COMPLEX THAN THAT OF THE AXON

• The initial excitement produced by the development of the ionic hypothesis was followed by a relatively quiet period in membrane biophysics that lasted from 1953 to 1965. This has now been followed by a period of ferment, a period that has changed our thinking about membrane channels. The Hodgkin-Huxley model demonstrated that the excitable properties of the axonal membrane could be explained merely by two gated channels and by leakage. The recent advances in biophysics make us realize that certain membranes, such as those of the cell body, are more complex. The new channels discovered in the cell body account for several interesting characteristics of the somatic membrane, including features that distinguish between identifiable cells, such as the capability for repetitive firing, the variety of endogenous firing rhythms, different degrees of accommodation, different types of afterpotentials, and synaptic potentials. In addition, some of these newly discovered channels are present in certain presynaptic terminals, where they are important in the modulation of transmitter release.

A corollary to these findings is a renewed appreciation that the axon membrane is fundamentally different in its signaling capability from those of the cell body and of the axon terminal. Whereas, from the perspective of signaling, the axon serves as a relay, conveying faithfully the impulses initiated in it, the cell body and the initial segment are the sites of synaptic integration, of impulse initiation, and of pacemaker activity. The presynaptic terminals, in turn, control transmitter release.

In looking back upon the recent period of discovery, three points of interest emerge. First, in almost all instances, the discovery of the new channel was first made in the nerve cells of gastropods. This, of course, is not surprising. Gastropod cells are large, accessible, and often identifiable, so the same cells can be examined repeatedly in different animals of the same species. Moreover, large cells allow insertion of multiple electrodes for voltage-clamping and for intracellular perfusion. As a result, one can study the membrane currents in these cells while changing the intracellular as well as the extracellular ionic environment. Finally, these cell bodies are usually free of synapses and dendrites and can therefore be isolated by ligation.

The second point is more surprising. None of the several ion channels found in gastropods turns out to be parochial. All of these channels now have been found in the central neurons of vertebrates. Moreover, all neurons so far examined—including the spinal motor neurons, the Purkinje cells of the cerebellum, the hippocampal pyramidal cells, and sympathetic ganglion cells—have at least some of these newly discovered channels. This speaks not only for the evolutionary conservatism of neuronal function and the universality of membrane properties, but also indicates that we are now on the verge of analyzing, in other mammalian cells, many properties of electrical excitability that previously had escaped attention.

Finally, studies with gastropod neurons as well as other nerve cells illustrate the importance of membrane biophysics for membrane biochemistry (a relationship stressed by Charles Stevens). The various ion channels are intrinsic membrane proteins. Any intelligent attempt to elucidate the molecular details of these proteins requires as a prerequisite an understanding of the biophysical properties of these channels. Without that knowledge, biochemical approaches inevitably would flounder. For example, without the realization that there are at least three quite different types of K^+ channels, an attempt to relate the biochemistry of the K^+ channel to its physiology would be confused and misguided.

THE NEW CURRENTS ARE NOW BEING RELATED TO NEURONAL EXCITABILITY, TRANSMITTER RELEASE, AND BEHAVIOR

• Progress in this area is perhaps as remarkable as any in the broad panorama of current neuroscience, although it is probably unfamiliar to many working neural scientists, certainly compared to the two-channel model of the axon. The unfamiliarity is perhaps due to the fact that the two-channel model of the axon emerged rapidly from the work of three collaborating scientists—Hodgkin, Huxley, and Katz. In contrast, the multichannel model of the cell body has emerged only gradually and has involved several independent investigators. The overriding reason for this meeting was, therefore, to attract the attention of other neurobiologists to this small, but nonetheless exciting revolution, a revolution that is important for molecular neurobiology, for synaptic transmission, and for behavior and learning.

REFERENCES

Anderson, C.R. and C.F. Stevens. 1973. Voltage clamp analysis of acetylcholine produced end-plate current fluctuations at frog neuromuscular junction. *J. Physiol. (Lond.)* **235:** 655.

Araki, T. and C.A. Terzuolo. 1962. Membrane currents and spinal motoneurons associated with the action potential and synaptic activity. *J. Neurophysiol.* **25:** 772.

Byrne, J.H., E. Shapiro, N. Dieringer, and J. Koester. 1979. Biophysical mechanisms contributing to inking behavior in *Aplysia*. *J. Neurophysiol.* **42:** 1233.

Carew, T.J. and E.R. Kandel. 1976. Two functional effects of decreased conductance EPSPs: Synaptic augmentation and increased electronic coupling. *Science* **192:** 150.

Carpenter, D.O. and B.O. Alving. 1968. A contribution of an electrogenic Na^+ pump to membrane potential in *Aplysia* neurons. *J. Gen. Physiol.* **52:** 1.

Castellucci, V.F., E.R. Kandel, J.H. Schwartz, F.D. Wilson, A.C. Nairn, and P. Greengard. 1980. Intracellular injection of the catalytic subunit of cyclic AMP-dependent protein kinase simulates facilitation of transmitter release underlying behavioral sensitization in *Aplysia*. *Proc. Natl. Acad. Sci.* **77:** (in press).

Cole, K.S. 1949. Dynamic electrical characteristics of the squid axon membrane. *Arch. Sci. Physiol.* **3:** 253.

Connor, J.A. and C.F. Stevens. 1971a. Voltage clamp studies of a transient outward membrane current in gastropod neural somata. *J. Physiol. (Lond.)* **213:** 21.

————. 1971b. Prediction of repetitive firing behavior from voltage clamp data on an isolated neuron soma. *J. Physiol. (Lond.)* **213:** 31.

Frank, K. and L. Tauc. 1964. Voltage-clamp studies of molluscan neuron membrane properties. In *The cellular function of membrane transport* (ed. J. Hoffman), p. 113. Prentice-Hall, Englewood Cliffs, New Jersey.

Frank, K., M.G.F. Fuortes, and P.G. Nelson. 1959. Voltage clamp of motoneuron soma. *Science* **130:** 38.

Geduldig, D. and D. Junge. 1968. Sodium and calcium components of action potentials in the *Aplysia* giant neurons. *J. Physiol. (Lond.)* **199:** 347.

Gerasimov, V.D., P.G. Kostyuk, and V.A. Maiskii. 1965. The influence of divalent cations on the electrical characteristics of membranes of giant neurons. *Biofizika* **10:** 447.

Gerschenfeld, H.M. and D. Paupardin-Tritsch. 1974a. Ionic mechanisms and receptor properties underlying the responses of molluscan neurones to 5-hydroxytryptamine. *J. Physiol. (Lond.)* **243:** 427.

————. 1974b. On the transmitter function of 5-hydroxytryptamine at excitatory and inhibitory monosynaptic junctions. *J. Physiol. (Lond.)* **243:** 457.

Gorman, A.F. and M.F. Marmor. 1970. Contributions of the sodium pump and ionic gradients to the membrane potential of a molluscan neuron. *J. Physiol. (Lond.)* **210:** 897.

Hagiwara, S., K. Kusano, and N. Saito. 1961. Membrane changes of *Onchidium* nerve cell in potassium-rich media. *J. Physiol. (Lond.)* **155:** 470.

Hodgkin, A.L. and A.F. Huxley. 1952a. Currents carried by sodium and potassium ions through the membrane of the giant axon of *Loligo. J. Physiol. (Lond.)* **116:** 449.

————. 1952b. The components of membrane conductance in the giant axon of *Loligo. J. Physiol. (Lond.)* **116:** 473.

————. 1952c. The dual effect of membrane potential on sodium conductance in the giant axon of *Loligo. J. Physiol. (Lond.)* **116:** 497.

————. 1952d. A quantitative description of membrane current and its application to conduction and excitation in nerve. *J. Physiol. (Lond.)* **117:** 500.

Hodgkin, A.L. and B. Katz. 1949. The effect of sodium ions on the electrical activity of the giant axon of the squid. *J. Physiol. (Lond.)* **108:** 37.

Hodgkin, A.L. and R.D. Keynes. 1955. Active transport of cations in giant axons from *Sepia* and *Loligo. J. Physiol. (Lond.)* **128:** 28.

Hodgkin, A.L., A.F. Huxley, and B. Katz. 1949. Ionic currents underlying activity in the giant axon of the squid. *Arch. Sci. Physiol.* **3:** 129.

————. 1952. Measurement of current-voltage relations in the membrane of the giant axon of *Loligo. J. Physiol. (Lond.)* **116:** 424.

Katz, B. and R. Miledi. 1969. Tetrodotoxin-resistant electric activity in presynaptic terminals. *J. Physiol. (Lond.)* **203:** 459.

Kerkut, G.A. and R.C. Thomas. 1964. The effect of anion injection and changes in the external potassium and chloride concentration on the reversal potentials of the IPSP and acetylcholine. *Comp. Biochem. Physiol.* **11:** 199.

Klein, M. and E.R. Kandel. 1978. Presynaptic modulation of voltage-dependent Ca^{++} current: Mechanism for behavioral sensitization in *Aplysia californica. Proc. Natl. Acad. Sci.* **75:** 3512.

————. 1980. Mechanism of calcium current modulation underlying presynaptic facilitation and behavioral sensitization in *Aplysia. Proc. Natl. Acad. Sci.* **77:** (in press).

Marmont, G. 1949. Studies on the axon membrane. I. A new method. *J. Cell. Comp. Physiol.* **34:** 351.

Meech, R.W. 1974. The sensitivity of *Helix aspersa* neurons to injected calcium ions. *J. Physiol. (Lond.)* **237:** 259.

Mudge, A.W., S.E. Leeman, and G.D. Fischbach. 1979. Enkephalin inhibits release of substance P from sensory neurons in culture and decreases action potential duration. *Proc. Natl. Acad. Sci.* **76:** 526.

Neher, E. 1971. Two fast transient current components during voltage clamp on snail neurons. *J. Gen. Physiol.* **58:** 36.

Neher, E. and H.D. Lux. 1971. Properties of somatic membrane patches of snail neurons under voltage clamp. *Pfluegers Archiv. Gesamte Physiol. Menschen Tiere* **322:** 35.

Pellmar, T.C. and W.A. Wilson. 1977. Unconventional serotonergic excitation in *Aplysia*. *Nature* **269:** 76.

Shapiro, E., V.F. Castellucci, and E.R. Kandel. 1980. Presynaptic inhibition in *Aplysia* involves a decrease in the Ca^{++} current of the presynaptic neuron. *Proc. Natl. Acad. Sci.* **77:** 1185.

Stevens, C.F. 1969. Voltage-clamp analysis of a repetitively firing neuron. In *Basic mechanisms of the epilepsies* (ed. H.H. Jasper et al.), p. 76. Little Brown, Boston.

Thomas, R.C. 1972. Electrogenic sodium pump in nerve and muscle cells. *Physiol. Rev.* **52:** 563.

Tillotson, D. 1979. Inactivation of calcium conductance depends on entry of Ca ions in molluscan neurons. *Proc. Natl. Acad. Sci.* **76:** 1497.

Weight, F.F. and J. Votava. 1970. Slow synaptic excitation in sympathetic ganglion cells: Evidence for synaptic inactivation of potassium conductance. *Science* **170:** 755.

Wilson, W.A. and H. Wachtel. 1978. Prolonged inhibition in burst firing neurons: Synaptic inactivation of the slow regenerative inward current. *Science* **202:** 772.

Ionic Channels in Neuromembranes: Methods for Studying Their Properties

CHARLES F. STEVENS

Department of Physiology
Yale University School of Medicine
New Haven, Connecticut 06510

• Communication among scientists, as among people in general, is often hampered by the vagaries of misinterpretation and definition of terms. Unfortunately, our primary tool of communication, technical terminology, can in itself be a formidable obstacle. This conflict is especially significant when modern developments in a broad field are presented for a wide audience.

A usual, and useful, solution to this problem is to include a glossary defining those terms and techniques least likely to be understood. This chapter is intended to serve the same function as a glossary, but by means of a different method. Instead of providing an alphabetical list of terms with their definitions, the words to be defined have been placed in the context of an essay on techniques for studying the properties of channels which must be understood if we are to know how the nervous system works. The chapter can be read straight through as an introduction to methods for studying channels, or it can be read piecemeal to find the meaning of particular terms. In the latter case, the reader should consult the Key to Terms given at the end of the book for an alphabetical index to the main words defined in this chapter.

CHANNEL PROPERTIES DETERMINE NEURON PROPERTIES

• Membrane proteins endow the otherwise rather featureless lipid bilayer envelopes of cells with properties that suit the cells for their function. The unique electrical properties of nerve cells arise from special classes of such membrane proteins, usually called **channels.**

These channels serve as the pathways for ions to cross the otherwise impermeant membrane. In addition, the channel proteins regulate ion flows: Channels can open or close either in response to a particular substance—a **neurotransmitter**—or by sensing the voltage across the nerve membrane. Ion flow through a channel is regulated by a process known as **gating**. Channels can be divided conveniently into two categories: **chemically gated** and **voltage gated.**

Within these two large categories, there are a substantial number of different channel types, although the exact number has not yet been determined. Classes of voltage-gated channels (those that have been distinguished in molluscan neurons are listed in Table 1) differ according to their sensitivity to the membrane electric field, the speed with which they open and close, and the particular ionic species that they permit to pass through their pores. Chemically gated channels differ as to the chemical nature of the ligand that is effective in opening and closing them, as well as the speed of opening and closing and the nature of ions that will pass through.

Just as different nerve cells have a variety of shapes, so do they contain different channel types: A class of channels abundant in one nerve cell's membrane may be absent completely in another. Furthermore, even those channels present are not distributed uniformly over the cell surface. Rather, certain channel types tend to be segregated in particular regions. Because the special electrical properties of nerve cells arise from the properties of the channels they contain, these differences in distribution and type of channel present give rise to classes of nerve cells with quite different characteristics. The richness of nervous-system function, then, depends not only on varieties of cell shape and interconnection in circuits, but just as importantly on which channels are present in the surface membrane and how they are distributed over that surface.

THREE CENTRAL PROBLEMS: PERMEATION, GATING, AND REGULATION

• To understand how the properties of channels express themselves in neuronal behavior, three problems must be solved. First, we must understand the physical mechanisms by which ions pass through channels—**permeation**—and particularly the process of **selectivity,** in which some ionic species are permitted to flow and others are excluded. Because the pores formed by integral membrane proteins are so small—they generally are less than $5 - 10$ Å at the narrowest point—the usual physical descriptions of ion movements in solutions are inapplicable. Special techniques, therefore, must be devised for describing ion movement within channels, and, at the same time, ion-protein interactions must be interpreted properly so that the mechanisms of selectivity become clear. To answer the first question, one would like to relate the movement and selectivity of ions to details of the molecular structure of the protein in question.

Second, the molecular mechanisms of **gating** must be investigated. Ligand binding or changes in the electric field induce a conformational change of the channel protein that serves to open or close the ion translocating pore. The goal here is to characterize this gating for each channel type and finally to understand its mechanisms at the molecular level.

Finally, the complicated mechanisms by which channels are distributed selectively over the membrane surface and their function is modulated must be understood. The question of **membrane protein regulation** and **metabolism** presumably is of general cell biological significance, and similar mechanisms probably are used to determine the distribution of all integral membrane proteins. There is an increasing body of evidence that the activity of channels may be modulated by chemical means in a rather long-term way. Mechanisms of **modulation,** which are thought to range from ligand binding to covalent modifications such as phosphorylation or methylation, must be investigated and the regulatory schemes involved must be discovered. The presentations in this book survey the progress that has been made regarding these three central questions through the use of a particular model, the nervous system of molluscs. Neuronal properties that arise from the presence of particular constellations of channel types are described, and these properties are related, at least in a preliminary way, to behavior.

CURRENT FLOW THROUGH CHANNELS IS AT THE HEART OF NEURONAL ELECTRICAL ACTIVITY

● Channels are responsible for the electrical activity of the nervous system; the flow of ions through the channels, producing an electric current, underlies this electrical activity. The simple equation that summarizes factors that influence current through a single class of channels

$$I = \gamma NF(V_m - E) \tag{1}$$

serves as a useful framework for discussing channel properties. In this equation, I is the current that flows through a population of channels; γ is the conductance of a single channel, a factor that reflects the permeability and selectivity properties of the channel; N is the density of channels (per unit membrane area) and depends on the metabolic and regulatory factors mentioned above; F represents

Table 1
Methods for Separating Currents

Channel Type	Designation or current	Synonyms	Blockers
Delayed K⁺	$I_{K(V)}$	delayed rectifier K⁺ current (I_K) $I_{K,V}$ I_K late	TEA[1] 4-AP[2] internal Ba⁺⁺
Early K⁺	I_A	$I_{K(A)}$ fast transient K⁺ current I_K early	TEA[1] 4-AP[2] I_A can be eliminated by holding V_m at −40 mV or more depolarized; I_A is completely inactivated in this voltage range
Ca⁺⁺-dependent K⁺	$I_{K(Ca)}$	C current (I_C)	I_{Ca} blockers eliminate $I_{K(Ca)}$ by blocking Ca⁺⁺ influx intracellular EGTA prevents the rise of internal Ca⁺⁺ and thereby prevents activation of this channel TEA[1] 4-AP[2] intracellular Ba⁺⁺ antagonizes Ca⁺⁺ activation of this channel
Na⁺	I_{Na}	fast inward current	TTX saxitoxin (STX) dose used can vary from nM (squid axon) to 10^{-4} M (some snail neurons)

| Ca++ | I_{Ca} | heavy metal ions: La^{3+}, Cd^{++}, Co^{++}, Ni^{++}, Mg^{++}—La^{3+} and Cd^{++} will block with normal $[Ca^{++}]_e$; the other ions are used primarily in conjunction with removal of Ca^{++} from the medium (approx. dose 30 mM) Ca^{++} removal from extracellular medium is completely effective only in conjunction with one of the above or with extracellular EGTA to buffer $[Ca^{++}]$ |
| | | organic blockers: D-600, verapamil, nifedipine (dose 10^{-4} M); these compounds have nonspecific effects on other channels |

---

[1] Internal TEA blocks I_K and I_A approximately equally. External TEA blocks $I_{K(V)}$ better than $I_{K(Ca)}$ in nudibranches but shows the reverse preference in *Aplysia*. It blocks $I_{K(V)}$ and $I_{K(Ca)}$ only at much higher concentrations.

[2] 4-AP blocks I_A with high affinity (10^{-3} M).

the fraction of channels in their open state and describes the gating of channels; and the quantity $V_m - E$ is the electrical driving force on ions moving through the channel, where V_m is the membrane potential and E is the reversal potential (that is, the voltage at which no net current flows through the channel). If only a single species of ion is available to pass through the channel, E is given by the Nernst equation

$$E = (RT/zF)\ln(C_o/C_i) \tag{2}$$

where z is the valence of the ion, C_i the inside ion concentration, C_o the outside concentration, and RT/F are thermodynamic constants. For monovalent ions at room temperature, the Nernst equation reduces, after conversion from natural into base-10 logarithms, to

$$E = 58 \log(C_o/C_i) \tag{3}$$

The quantity $g_m = \gamma NF$ is known as the **membrane conductance** for a particular class of channels. If the three quantities that combine to make up the g_m for a given type of channel are understood, we have a complete characterization of that channel's behavior. The meaning of factors contributing to g_m will be considered in more detail below, as will the techniques for studying them.

THE VOLTAGE CLAMP SIMPLIFIES THE STUDY OF CHANNELS

• Many modern studies of channels employ an electronic apparatus known as the **voltage clamp.** Although several forms of voltage clamps are available, they all use negative feedback to keep the voltage constant across the cell membrane and they all provide some means for measuring the current necessary to do this. Voltage clamps are used for two main reasons. First, the quantities that characterize channels in equation 1—the single channel conductance, γ, the fraction of open channels, F, and the density of channels, N—all are directly proportional to membrane current (I_m) when voltage, V, is kept constant, but not otherwise. It is most convenient, then, to measure current at a constant voltage because the quantity measured then bears a simple relationship to channel characteristics such as the gating function, F. A second, and often more important, reason for using a voltage clamp is that the behavior of many channels is influenced by the voltage across the membrane containing them. For example, voltage-gated channels are opened by positive-going shifts in membrane potential (V_m). When channels open, current flows, and this current tends to change the voltage. If voltage is not controlled, channel opening influences itself through V_m changes, and analysis of gating becomes hopelessly complex. For both of these reasons, nearly all studies of channel properties make use of a voltage clamp of some sort.

All voltage clamps work by negative feedback. The actual voltage across the cell membrane is measured and the difference between the desired voltage—known as the **command**—and the actual voltage is amplified and fed back with a sign that diminishes the difference. The electronic apparatus that amplifies the **error signal** (that is, the difference between the actual and the desired voltage) is known as the **voltage-clamp amplifier** and is simply a high-gain and high-frequency response differential amplifier that can supply currents of the required magnitude. All negative feedback systems of this type will become unstable and oscillate if the gain of the feedback amplifier is too high, so that the maximum **voltage-clamp gain** permitted is limited by the requirement that the feedback system be stable.

Voltage Clamps Are Imperfect

Although the idea behind the voltage clamp is a simple one, the technical realization of this apparatus can be very difficult indeed. In fact, no voltage clamp is perfect, and an understanding of the limitations in this method is important for a proper interpretation of data. The following paragraphs define some terms associated with voltage-clamp use and discuss a few of the important limitations of the method.

The whole point of the voltage clamp is that voltage be maintained constant so that current reflects, for example, the time course of channel opening and closing. It turns out that, in fact, voltage is not held at precisely the command level for several reasons. Three important sources of error must be considered. First, the voltage clamp is completely free of error only when the gain of its amplifier is infinite. For practical amplifiers, errors range from less than 0.1% up to 10–20%, depending on the exact circumstances.

Second, the **series resistance** artifact is important. The physically relevant quantity, V_m, in equation 1 is the voltage difference between the inside and outside surfaces of the membrane. In experiments, however, voltage differences are actually measured between an electrode inside the cell and a relatively distant ground or reference electrode in the bath containing the preparation. Because the recording electrodes are not exactly at the surface of the membrane, a volume of fluid between the membrane surface and the voltage-measuring electrode gives rise to a small resistance in series with the membrane (the series resistance). Any currents that flow through the membrane also flow through this resistance, and because currents flowing through a resistor give rise to voltage differences, the voltage at the membrane surface is not identical with that measured by the electronic apparatus. Furthermore, the magnitude of this error in measuring voltage increases as the current flow through the membrane increases. The voltage-clamp apparatus can only control the measured voltage and not the actual voltage just at the membrane surface. Consequently, the actual voltage deviates to some extent from the measured voltage, and the extent of the deviation depends on the magnitude of currents. Errors caused by this deviation are known as **series resistance artifacts.**

A final source of voltage error arises when, as usually is the case, the entire interior of the cell being studied does not have the same voltage; such differences in internal voltage arise from internal current flows and the resistance of cytoplasm. A technique that insures a cell will have the same voltage throughout is known as a **space clamp,** and failures of such isopotentiality are known as lack of space clamp. Because the currents generally measured flow through the entire surface membrane of the cell, a failure of spatial voltage uniformity gives rise to errors because bits of membrane with different voltages contribute to the total measured current.

Three Different Types of Voltage Clamps Are Used

Three different types of voltage-clamp arrangements are currently in common use. In some instances, a single electrode can be used to record voltage and pass current simultaneously. In such a case, we have the **single-electrode voltage clamp.** This is an easy method when the electrode has a very low resistance, such as is the case with an axial wire in a squid axon; but the situation becomes more difficult if a microelectrode is used. Because current flow causes voltage drops across a single microelectrode, special electronic circuitry must be used if the same electrode is to control V_m rapidly by passing current and also simultaneously record that V_m. Essentially, the electronic apparatus makes a single electrode into two electrodes by rapidly switching in time between current passing and voltage recording.

The most usual voltage-clamp method for nerve cells is the **two-electrode clamp,** by which one electrode records voltage and a second is used for passing the current that maintains that voltage constant. This arrangement is also known as the **point clamp** because only the point in the cell where the voltage-recording electrode resides is maintained definitely at the command voltage. If the cell under investigation fails to have a spatially uniform voltage, then errors can arise. These errors can be insignificant if the investigator is interested in changes in permeability that occur only close to the voltage-recording electrode. In this situation, changes in current accurately reflect changes in permeability with the voltage being fixed and known. The least favorable situation for the two-electrode voltage clamp is a long, cylindrical structure, such as an axon or muscle fiber, in which the investigator wishes to study permeability changes that occur over the entire cell. Because of voltage differences along such a cell, a two-electrode clamp is useless.

The electrical properties of long cylinders happen to make possible another kind of voltage clamp, the three-electrode arrangement. In the **three-electrode voltage clamp,** microelectrodes are inserted into the elongated cell with specific spacings relative to its end, and current is measured from the voltage difference between two of these electrodes. Under many circumstances, the results obtained in this situation can be shown to be the same as would have been gotten from a two-electrode voltage clamp in the same cell with a uniform internal potential.

Voltage Clamps Sometimes Cannot Change Voltage Rapidly Enough

As will be described in more detail in a later section, the voltage, V_m, appearing in equation 1 must be changed quickly from one level to another to study the opening and closing of voltage-gated channels. Two sources of difficulty are associated with such rapid changes in voltage (**voltage steps**). The first difficulty is that, because of limitations in the speed of voltage-clamp apparatus, the time of transition from one voltage level to another can be comparable to relaxation times of gating processes being studied. For example, even if the transition of a command pulse between −50 mV and 0 mV took only 1−2 μsec, the actual V_m might require as much as 1−2 msec to settle at the desired level. If the investigator were studying very rapid channels (like Na^+ channels) whose opening and closing time constants are on the order of 1 msec, voltage appearing in equation 1 would be varying while channels were opening, so that the interpretation of the measured currents would be quite complicated. The precise speed with which voltage-clamp apparatus can cause a transition from one voltage level to another depends on the details of the situation. The usual **settling time constant** for a two-microelectrode voltage clamp is approximately the electrode resistance multiplied by the cell membrane capacitance divided by the gain of the voltage-clamp amplifier. The voltage-clamp amplifier gain is usually around 1000, its upper limit being set by the stability of the feedback system.

The second problem associated with transitions in voltage arises because cell membranes have a very considerable capacitance, and the **displacement current** that flows through a capacitor is proportional to the capacitance size and the rate at which voltage across the capacitor is changing. This means that during the time the voltage across a cell membrane is changing, a capacity current, known as the **capacity transient,** is added to the simple ionic currents that result from permeability changes (see equation 1). The duration of the capacity transient ranges from a few tens of microseconds to several milliseconds and generally corresponds to the length of time it takes for the voltage transient to reach its final value after a step change in V_m. Capacity currents, then, obscure the underlying ionic currents that we wish to study.

LEAKAGE AND CAPACITY CURRENTS CAN OBSCURE MORE INTERESTING PHENOMENA

• Even in the resting state, nerve cell membranes have at least a slight permeability to ions. The resting resistance of neurons is termed the **leakage resistance** and generally behaves in a relatively simple way in that it follows Ohm's law. Currents that flow through this leakage resistance are usually uninteresting and tend to obscure the processes under investigation. Typically, then, the leakage currents, which are proportional to voltage change, are subtracted from the total observed currents; this subtraction process is known as **leak correction** or **leakage subtraction.** Because the capacity transient behaves in a linear fashion, the same sort of subtraction procedure can be used to eliminate the capacity transient from records, and this is done frequently. Even though the capacity transient can be subtracted successfully, it is important to remember that its time course reflects a period of nonconstant voltage, so ionic currents that flow during this time must be interpreted with care.

GATING DEPENDS ON VOLTAGE OR AGONIST CONCENTRATION, AND ON TIME

• Channels can, as stressed earlier, be divided into two categories: chemically gated and voltage gated. The chemically gated channels are opened by specific neurotransmitters and other chemicals, and the voltage-gated channels respond, by opening, to the voltage difference across the nerve cell membrane. These two categories are not necessarily mutually exclusive, and it is well known, for example, that the opening and closing of certain chemically gated channels is influenced by V_m. Nevertheless, the distinction is a useful one and the two categories of gating mechanisms require different techniques for their study.

Gating is represented in equation 1 by the function F, which specifies the fraction of channels that are in the open state. Here, the tacit assumption has been made that a channel is either open or closed, and no intermediate conducting states exist. The gating function F depends on voltage for the voltage-gated channels and on the concentration of some agonist for the chemically gated channels. In addition, few channels open and close instantaneously, so the function F generally depends also on time (t). Gating then must be specified by the relationship between fraction of channels open and either voltage or concentration and also by a **relaxation time constant** that describes the length of time it takes for a population of channels to reach their steady fraction open. For voltage-gated channels, the relationship between fraction open and voltage is called the **activation curve,** and for chemically gated channels, the analogous relation between fraction open and agonist concentration is known as the **dose-response curve.** The relaxation time constants themselves depend upon voltage or agonist concentration, and a full characterization of gating usually requires that time-constant vs voltage or time-constant vs concentration relations be investigated. In some instances, relaxations follow a simple exponential time course, but, more typically, voltage-gated channels open and close with more complicated time courses that involve lags and delays. Thus, although the simplest channels might be characterized completely by their activation curve and time-constant curve, most channels require additional information for a full description of their properties.

To study voltage-gated behavior, it is necessary to subject the membrane to step changes in V_m and then to measure the currents that flow. A graph of steady-state current as a function of voltage is called the **steady-state I-V curve,** and this I-V relationship is used to determine the activation curve (see equation 1). Two methods are used for measuring relaxation time constants, one for the voltage range over which channels are activated and the other for hyperpolarized voltages where all channels are closed. For the first method, voltage is stepped from the resting level (where channels normally are closed) to a more positive value. The relaxation time constants then are determined from the approach to the steady level of activation. For the voltage range where channels are closed, the second method, using **tail currents** (I_t), is required. Voltage is first stepped to a value where many channels open and then to a more negative level. Immediately after the step to a negative voltage, channels are caught in the open state and they then close over time. Currents that reflect the decreased number of open channels, I_t, yield a relaxation time constant for the negative voltage. Combining these methods, one can measure relaxation time constants over a wide voltage range.

Tail currents have a second use: They can provide an estimate for the reversal potential, E, when it happens (as for K^+ channels) to fall in a voltage range below which channels are activated. When $V_m = E$, the driving force on currents vanishes in equation 1, so E can be determined by finding the value of V_m at which a class of channel's I_t disappear.

Chemically gated channels are studied by maintaining voltage constant and applying different agonist concentrations. In some instances, the chemical composition of the bathing medium cannot be changed rapidly enough to study the actual time course of channel opening and closing, and, in these instances, the kinetics of chemically gated channels must be studied with single-channel recordings or fluctuation analysis (methods described later).

TO STUDY A PARTICULAR CHANNEL TYPE, CURRENT THROUGH OTHER CHANNELS MUST BE ELIMINATED

• As indicated earlier, equation 1 applies to a single class of channels, but the nerve membrane characteristically contains an intermixed population of several or many classes. As a result, the currents from various classes of channels add together and complicate the interpretation. To characterize a particular channel class, it is necessary to perform a **separation of currents,** that is, to eliminate, by one or more of these methods, all currents but the ones flowing through channels being investigated. In some instances, separation of currents can be achieved by exploiting inherent characteristics of the membrane channels being studied. For example, one class of channels may activate between −60 mV and −40 mV, whereas a second class of channels may not activate until above −40 mV. In this case, the first class of channels could be studied without contamination from currents of the second class, but the reverse would not hold. If two classes of channels have very different relaxation time constants, separation can be achieved simply by ignoring very early or very late times.

Since different channels tend to conduct different ions, sometimes currents through a particular class can be eliminated by removing the ions that pass through. For example, currents through Na^+ channels are sometimes eliminated successfully by replacing all Na^+ in the bathing medium with an impermeant cation such as TEA^+. One difficulty with using ion substitution as a means for eliminating currents through a particular channel is that ionic selectivity of channels is seldom absolute; in addition, it can be difficult to find a completely impermeant species. Another difficulty is that ion substitutions frequently can have other pharmacological effects on the remaining channels.

The third important class of separation techniques depends upon using pharmacological blocking agents. Certain toxins and other

chemicals can block particular varieties of channels selectively, sometimes with great specificity, and this technique is standard for eliminating certain classes of currents. A list of pharmacological agents
used in separation appears in Table 1.

CHANNELS RECTIFY

• The quantity γ in equation 1 reflects the permeability properties of
a single open channel and has a numerical value that specifies the
conductance of the channel (reciprocal of the resistance for one
channel). In some situations, an open channel behaves like a simple
ohmic resistor over the range of voltages investigated, which means
the value of γ is constant. Typical values for the single-channel
conductance seem to range from several pS to 50 pS (1 pS corresponds to a conductance of 10^{-12} reciprocal ohms, or Siemens). Because
current is carried through channels by the movement of specific ions,
it is obvious that the value for γ will depend upon the concentration
of permeant ions; if no charge carriers are available, no current can
flow, and the channel will have a very low conductance. In all cases,
then, the value of γ depends upon ion concentrations. It is perhaps
less obvious that this value also can, and frequently does, depend
upon the voltage across a membrane driving ions into the channel. In
these instances, the channel is said to exhibit **rectification,** which
means currents of some directions or magnitudes require less voltage
driving force than do others. The single-channel conductance γ depends not only on ion concentrations, then, but also on voltage, V.
   The fact that the value of γ can depend on voltage complicates
the analysis of channels because both γ and F can, for voltage-
sensitive channels, combine to influence the magnitude of the current
that flows at a given voltage. To distinguish influences of γ from
those of F, the value of γ for each voltage is measured. This measurement, known as constructing the **instantaneous I-V curve,** depends
on the fact that γ changes its value instantaneously on the experimental time scale, whereas the opening and closing of the channels
(reflected in the value of F) takes a relatively longer time. To determine the dependence of γ on voltage, the experimenter causes the
voltage to step from one level to various other levels and observes the
magnitude of current that flows just after the step has occurred and
before there has been time for the fraction of channels open (F) to
change. Once the instantaneous I-V curve has been determined in this
way, it is possible to separate out influences of rectification from
those of the changing number of open channels, F.
   The laws governing the movement of ions through channels are,
of course, interesting in their own right, and the rectification properties of channels provide one way of studying ion permeation. Al-

though the physical basis for the rectification is not yet understood in complete detail, it often is convenient to have an equation that describes the instantaneous *I-V* curve. The **Goldman-Hodgkin-Katz (GHK) equation** generally is used for this purpose. The GHK equation is derived from physical principles that boil down to a combination of Ohm's law and Fick's law of diffusion. This combination is called the **flux equation,** and by solving the flux equation with the assumption that voltage changes through a pore are linear (the so-called **constant-field assumption**), one can obtain the GHK equation. The GHK equation, sometimes also known as the **constant-field equation,** is not very accurate in most instances, but it does give a conventional way for treating rectification. This equation also predicts that the degree of rectification changes when ionic concentrations are altered, and therefore the GHK equation provides a quantitative way of dealing with concentration effects of γ. Furthermore, the GHK equation contains quantities, known as **permeabilities,** that can be different for different ions. For example, a Na^+ channel may have permeabilities for Na^+ that are 10 to 20 times greater than those for K^+. The GHK equation gives a quantitative way of measuring such permeability ratios. In summary, the GHK equation offers a convenient characterization of the quantity γ and provides an important framework for thinking about ionic selectivity and permeability of channels; the equation is not, however, very accurate over the full range of voltages used in experiments.

More recent treatments of permeation and selectivity have started from a view that ionic motion within a channel is best viewed as a series of hops—this is like ionic diffusion in crystals—whose rate (according to **Eyring rate theory**) decreases exponentially with the highest of the energy barriers separating two ion positions (or, more generally, two states). This approach, using **ion-hopping models,** is replacing the traditional treatment of the channel interior as a continuous medium and is more realistic because of the small dimensions of pore sizes encountered (on the order of 5 Å). Rectification, as described by *I-V* relationships for a single pore, has been derived for models incorporating this more modern view. Treatments of channels in terms of ion hopping have not replaced the GHK approach because the resulting equations tend to be more complicated than the GHK equation.

CHARGES ON THE MEMBRANE SURFACE INFLUENCE THE BEHAVIOR OF CHANNELS

• Oftentimes one wishes to change the composition of the bathing medium to study channel properties. Such changes in composition can have a substantial influence on the properties of channels because of charged groups fixed on the membrane surface. The phenomena associated with these charges are known as **surface-charge effects** and fall into two categories. The first category relates to the direct effect that surface charges have on the molecular voltage sensors responsible for gating. Fixed surface charges contribute directly to the electric field experienced by proteins embedded in the membrane, so alterations in surface charge can influence these proteins without changing the voltage across the membrane that the experimentalist measures. Altering the ionic composition of the bathing medium, particularly changing the concentration of divalent ions, can influence the effective surface charge and thereby cause the equivalent of V_m changes that are not detected by the voltage-measuring electrodes. Such changes in effective V_m of course alter the behavior of voltage-gated channels.

The second influence of surface charges has to do with the **surface concentration** of various ions. In the presence of surface charges, neutralizing ions with a charge opposite that fixed on the membrane surface must compete for a place near the membrane. When negative charges are present on the membrane surface (as they typically are), divalent cations such as Ca^{++} can displace, for example, Na^+ from near the membrane surface and decrease Na^+ availability for channels. Thus, by this displacement mechanism, changing the $[Ca^{++}]$ can influence the degree of rectification exhibited by a Na^+ channel. Surface-charge effects are often an important complication in experiments that involve changes in ion concentrations, particularly of divalent ions.

SOME IONS CAN BLOCK SOME CHANNELS

• Often a particular ionic species will either pass through a channel or will be excluded and inert. In some instances, however, organic or metal ions can enter a channel but cannot fit through the narrowest part; they act like plugs. This mechanism is specific for channel type and ion species and is thought to account for the blocking of potassium current (I_K) by TEA and also for certain local anesthetic effects in both nerve and synaptic membranes. Blocking is important as a pharmacological tool in separating currents, is a notable source of artifacts when one inadvertently includes a blocking ion, and serves as a tool to probe the characteristics of channels (by varying the size and chemical characteristics of the blocking ion). In some instances, the blocking effect occurs very rapidly on the normal experimental time scale and so appears to occur instantly. In other circumstances, however, blocking and unblocking of channels occurs relatively slowly, so that blocking effects can look like channel closing. In this latter case, the danger is that the time course of blocking will be confused with characteristics of the channel under investigation.

COMPOSITION OF THE INTRACELLULAR MEDIUM CAN BE CONTROLLED

• In isolated preparations, such as molluscan ganglia, the composition of the bathing medium is controlled easily. The intracellular composition, however, reflects a variety of factors, including extracellular concentrations of small molecules, the permeability of the surface membrane to ions and other substances, and the activity of cotransport systems and metabolically driven pumps. A nonconstant or uncertain internal environment is clearly a complicating factor in experimental analysis. For certain types of experiments, control of intracellular concentrations can be crucial. Over the past several years, a number of laboratories have developed methods for changing the composition of internal fluids by **internal perfusion** or **dialysis.** The most commonly used technique is to suck a neuron soma, usually a molluscan nerve cell body, onto the tip of a glass electrode and to rupture the membrane in the aperture of the electrode. Under favorable circumstances, the nerve membrane surrounding the breached neuron surface adheres closely to the electrode tip and forms a high-resistance seal. Usually, a fine wire protrudes from the tip—this can be used to rupture the membrane—and serves to decrease access resistance to the cell's interior, so that a relatively simple single-electrode voltage-clamp arrangement can be used for studying channels. Over a period of 5–20 min, depending on the exact circumstances, small molecules in the cytoplasm exchange with those in the electrode tip, so that by changing the composition of fluid in the electrode, one can control the intracellular composition of the cell. So far, experimentalists have concentrated on changing the internal ionic composition of the nerve cell studied by this method, but the concentration of organic compounds, such as cyclic nucleotides and enzymes, can also be controlled, and this will be exploited in the future. A possibility even exists for using these cell sacs, which retain their nucleus and probably other large organelles, for biochemical and membrane-flux studies.

In some instances, it is impossible or inconvenient to use the internal perfusion technique just described but it is still desirable to be able to change the ionic composition of the cytoplasm. This can be achieved by temporarily inserting ionophores, such as nystatin for changing $[K^+]$ or A23187 for permitting a Ca^{++} influx, into the surface medium and then appropriately regulating the extracellular concentration of the ions one wishes to change. The ionophores can, if necessary, be washed out, and the cells can be studied with high-resistance membranes relatively impermeant to the species with which they have been loaded or depleted.

FLUCTUATION ANALYSIS: WE CAN LEARN FROM NOISE

• The opening and closing of channels are both random events. Even when the average number of channels open is fixed, the exact number open fluctuates from instant to instant around the mean. These fluctuations in number of channels open give rise to variations in membrane current, which is often called **noise.** Because the same physical processes determine both the statistical structure of noise and the more usually studied average membrane responses, an appropriate analysis of these statistical fluctuations can provide additional information about channel characteristics. The investigation of statistical structure in membrane electrical noise is termed **fluctuation analysis.**

Fluctuation analysis is useful in two main circumstances. The first relates to the properties of chemically gated channels. Often, it is impossible to change the concentration of neurotransmitters rapidly enough to permit the gating characteristics of chemically gated channels to be investigated. Quantities such as the mean length of time a channel stays open can, however, be determined by analysis of fluctuations observed with a constant dose of agonist.

The second class of uses involves investigations of the physical mechanisms that underlie gating. It can happen that distinct physical mechanisms will give rise to identical average behavior for a population of channels. In this case, ordinary voltage-clamp experiments cannot distinguish between the two possible mechanisms. In general, however, different mechanisms—even those producing the same average behavior—yield differences in the statistical structure of fluctuations. By studying fluctuations, one can decide which of several alternative mechanisms is the one used by nerve cells.

THE STATISTICAL STRUCTURE OF FLUCTUATIONS IS CHARACTERIZED BY THE SPECTRUM

• The amplitude of any fluctuating process is usually specified by the standard deviation, or the square of this quantity, the **variance.** The variance is simply the average squared deviation from the mean value of a fluctuating quantity. Fluctuations, however, occur at various rates: Some random processes vary slowly, whereas others vary rapidly. The variance thus gives an incomplete characterization of a random process in that it does not give any indication about the speed with which fluctuations are occurring, and rapidity is an extremely important character of noise. The standard method for characterizing the rapidity of fluctuations is to calculate the variance for narrow frequency bands over a wide range of frequencies. In effect, the fluctuations are filtered to remove all frequency components except those at 1 Hz (± ½ Hz), and the variance for these reduced fluctuations is computed. The process is repeated again for 2 Hz (± ½ Hz), then for 3 Hz (± ½ Hz), etc. A graphic representation of variance as a function of frequency, plotted on a double logarithmic scale, is called a **spectrum.** The spectrum for slow fluctuations would have appreciable amplitudes only for the low frequencies, whereas a rapid process would be characterized by a spectrum with appreciable amplitudes to quite high frequencies. The spectrum is useful in characterizing a random process because, although the exact path of the fluctuations is almost always different for two samples, the spectrum is (within sampling errors) the same from sample to sample.

Probability theory tells us how to predict the spectrum that would arise from any particular mechanism. For example, if a population of identical, independent channels has a constant probability per unit time of opening and, once open, a constant (in general, a different constant) probability per unit time of closing, the fluctuations produced are characterized by a **single-time-constant spectrum.** This spectrum is often, and incorrectly, called a **Lorentzian** by analogy to the similar, but not identical, Lorentz function that arises in connection with nuclear magnetic resonance studies.

The single-time-constant spectrum associated with the simple open-shut channel behavior just described has the form

$$S(f) = 4\gamma g_m \tau / [1 + (2\pi f \tau)^2] \tag{4}$$

where $S(f)$ is the spectral density (variance contributed by a 1-Hz band width around the frequency f), g_m is the mean membrane conductance, γ is the single-channel conductance, and τ is the mean length of time the channel stays open. A characteristic frequency, f_o, known as the **corner** or **cutoff frequency,** is defined by $f_o = 1/2\pi\tau$ and is that frequency for which $S(f)$ is reduced to one-half the limiting value the spectrum attains for low frequencies. For this simple kind of channel, we can measure both single-channel conductance and mean open time from fluctuation analysis. Furthermore, comparison of the theoretical and predicted spectra provides one check on the validity of the postulated gating mechanism. Equation 4 should fit the experimentally determined spectrum for a simple open-shut channel. More complicated channel mechanisms produce more complex spectra, but single-channel conductance and open lifetime typically are reflected in the form of the spectrum.

SINGLE CHANNELS CAN BE STUDIED IN ISOLATION

• In certain favorable circumstances, currents through a single channel can be recorded. To make a **single-channel recording,** the experimenter uses a glass electrode with a pore diameter in the range of less than 1 μm to several micrometers. This electrode is pressed onto the clean surface of a cell containing channels so that a seal is formed between the glass and the membrane surface. Current through the membrane patch is then recorded. Channels are made to open either by changing V_m or by having an agonist within the recording electrode. With proper attention to detail, this arrangement makes possible resolution that is adequate to detect the several-picoamp currents that attend channel opening. For simple systems, single-channel current recording gives about the same information as does fluctuation analysis. In more complicated situations, however, single-channel studies can, when they are possible, provide important advantages.

THREE TECHNIQUES FOR STUDYING RESTRICTED REGIONS OF THE NEURON SURFACE

• As was emphasized earlier, channels are not distributed uniformly over the entire surface of the neuron, but rather certain channel classes are restricted to particular regions of the membrane. This channel distribution, of course, has important implications for nerve cell function. Three techniques are available for studying restricted regions of the cell membrane. First, for chemically gated channels, the appropriate agonist may be applied to only a small region of the membrane. This is done typically by iontophoresis of the agonist from an electrode placed very close to the cell surface, even though pressure ejection of material from fine micropipettes can achieve fairly good localization. Because the agonist is applied to only a small part of the membrane surface, changes in the total currents measured with the voltage clamp reflect the response of the channels only in the excited membrane region.

A second method, applicable to voltage-gated channels, is **patch recording.** For this technique, the arrangement is similar to that used for single-channel recording except that the aperture of the glass electrode is somewhat larger so that currents are collected from a rather large population of channels. Usually, the cell is voltage-clamped with a conventional two-electrode system, and currents are collected from the small region under investigation by a patch electrode. The two main difficulties with the patch-electrode technique arise from the presence of channels underneath the walls of the patch electrode and the loss of currents through an imperfect seal between the glass recording electrode and the cell membrane.

Close to the cell surface, currents flow almost perpendicularly to the membrane. These currents are associated with a voltage drop across the resistance of the bathing medium, so that by measuring voltage differences between closely spaced points, membrane currents can be estimated. This provides a third method for studying a limited membrane region. One way to measure such local currents is to mount a pair of electrodes close together (a **differential pair**) and record the voltage difference between them when they are placed next to the cell surface with proper orientation. Another technique is to use a single electrode that is moved rapidly back and forth over a small distance (a **vibrating probe**). A vibrating probe has very high sensitivity, but it cannot be used to study rapidly varying processes because estimates of membrane current are made by subtracting values obtained at the successive positions of the electrode. The relaxation time constants of the process being studied must be long compared with the vibration period of the probe, which is limited, for technical reasons, to frequencies in the auditory range. Both the differential pair and the vibrating probe can be used to study either voltage-gated or chemically gated channels.

INTRACELLULAR Ca^{++} CONCENTRATION ($[Ca^{++}]_i$) CAN BE MEASURED BY TWO METHODS

• $[Ca^{++}]_i$ is highly regulated in all cells, and many cells, including neurons, use changes in $[Ca^{++}]_i$ for signaling and regulatory functions. To investigate these processes, internal $[Ca^{++}]$ must be measured. Two classes of techniques are available for measuring the intracellular concentrations of this important ion. First, **ion-sensitive electrodes** can be fabricated that respond with a voltage that is related to the $[Ca^{++}]$ at their tips. These electrodes work for Ca^{++} much the way a pH electrode does for H^+.

The second, more commonly used method employs indicator dyes that report $[Ca^{++}]_i$. Fluorescence of the naturally occurring protein **aequorin** is related jointly to $[Ca^{++}]$ and the quantity of aequorin available, so this protein can be used as a sensitive measure for changes in low $[Ca^{++}]$. The absorbance of light of certain wave lengths by the dye **arsenazo** depends on $[Ca^{++}]$, and this dye also can be used as a Ca^{++} indicator. Other dyes also respond to Ca^{++}, but they either are not readily available or have characteristics that make them less favorable than arsenazo. Whenever a dye is used, sufficient quantities must be injected into the cell, nonlinearities of the concentration-response curve must be taken into account, and it must be realized that local $[Ca^{++}]$, for example, at the membrane surface may be physiologically important, whereas the measurement typically reflects concentration changes over a much larger region of the cell.

FUTURE DIRECTIONS

• Some of the techniques described in this chapter are very new and still under development, whereas others have been standard for 5 or 10 years. A number of years will be needed to work out channel properties with the methods already available and with those (like single-channel recording) that are emerging, but certain directions for future developments are already clear. The next large classes of problems will involve understanding regulation of channels and their metabolism and the relationship between molecular structure and function. Both of these directions doubtless will require additional techniques. In the first case, single-cell perfusion and biochemical techniques will have to be perfected and data using these methods will have to be correlated with voltage-clamp analysis. For the second class of problems, we will need to refine biochemical studies of integral membrane proteins, and particularly to discover ways of getting structural information other than with classical sequence analysis and X-ray crystallography. As these new methods are developed, we can anticipate ultimately approaching the goal of understanding behavior at the molecular level.

General Considerations in the Study of the Ca^{++} Channel

Based on a presentation by

SUSUMU HAGIWARA

Department of Physiology
UCLA School of Medicine
Los Angeles, California 90024

• It is now clear that Ca^{++} channels are distributed among a wide variety of excitable tissues of different animals. In general, Na$^+$ action potentials are found wherever the membrane is specialized for impulse conduction. Action potentials with prominent Ca^{++} components, on the other hand, are more common in tissues where the impulse acts to trigger an effector process, such as contraction, secretion, bioluminescence, etc. There are many tissues in which the biological significance of Ca^{++} spikes is not yet clear. The soma of the molluscan neuron is one.

Hagiwara's earliest studies on the soma of the molluscan neuron, begun in 1958, were focussed on determining whether or not voltage-clamp experiments could be done with two intracellular glass micropipettes. After an intensive search, the marine pulmonate *Onchidium* was chosen as the test organism having the most promising nerve cells for such a study. In the initial voltage-clamp experiments with this preparation, it was assumed that the inward current evoked by depolarization was carried by Na$^+$, in keeping with the Hodgkin-Huxley theory of the squid axon. However, Hagiwara decided to test this point by superfusing the ganglion with seawater in which all NaCl had been replaced by sucrose. One hour of this perfusion reduced, but did not eliminate, the inward current, and the reduction was reversible. The conclusion that the remaining inward current was carried by Ca^{++} appeared premature, and therefore this current was described as an inward surge, without the ion species involved being specified (Hagiwara et al. 1961). The actual assertion that divalent ions carry the inward current in molluscan nerve cells was first made by Oomura et al. (1961).

The report of this presentation was prepared by G. Yellen.

The ultimate goal of biophysical studies of I_{Ca} is to define the properties of the channels that carry I_{Ca}. Of particular interest is the mechanism of permeation of ions through the Ca^{++} channel: To what extent is the permeation process concentration-dependent? How selective is the channel for Ca^{++}? What is the conductance of a single channel? In addition, the characteristics of the Ca^{++}-channel gating mechanism—voltage sensitivity, kinetics, and mechanisms of activation and inactivation processes—need to be determined. These questions are taken up by Brown, Tillotson, and Llinás (all this volume).

The following more-general questions are considered here: (1) What criteria must be met to demonstrate the presence of Ca^{++} channels? (2) Is there more than one species of Ca^{++} channel? (3) What special experimental problems arise in the study of Ca^{++} channels? (For reviews, see Hagiwara 1973; Hagiwara and Byerly 1981.)

IDENTIFICATION OF Ca^{++} CHANNELS

• There are four phenomenological criteria that can be used to determine whether the action potential in a particular tissue is generated by Ca^{++} influx. By measuring inward current, rather than the height of the action potential, these criteria can also be applied in voltage-clamp experiments to determine whether the inward current is carried by Ca^{++}.

1. The overshoot of the action potential should increase with an increase of $[Ca^{++}]_e$. When $[Ca^{++}]_e$ is in the low range, the overshoot tends to increase with the Nernst slope for Ca^{++}, but the overshoot usually levels off as $[Ca^{++}]_e$ is increased, because the Ca^{++} channel is a saturable system.
2. Action potentials should persist in Na^+-free media if the external solution contains either Ca^{++}, Sr^{++}, or Ba^{++}.
3. The action potential should be blocked by Ni^{++}, Cd^{++}, Zn^{++}, Co^{++}, or La^{3+}, or by organic Ca^{++} antagonists such as verapamil, D-600, and nifedipine, but not by TTX. Since none of these Ca^{++} blockers are as specific for Ca^{++} channels as TTX is for Na^+ channels, they also may block Na^+ and K^+ channels when their concentrations are too high.
4. Ca^{++} influx during the action potential should be sufficient to account for the charge transfer during the rising phase of the spike. This test has been performed in only a few cases.

IS THERE ONLY ONE KIND OF Ca^{++} CHANNEL?

• Whether the Ca^{++} channel is the same in all tissues in which it has been examined must be known if generalizations are to be made about its properties. It is generally accepted that the Na^+ channel has a unique molecular structure that is highly conserved across many species. Such a uniformity of structure is less likely for the Ca^{++} channel. It now seems more likely that several different kinds of Ca^{++} channels exist. This conclusion is based on the large differences observed in kinetics, pharmacological sensitivity, ion selectivity, and mode of activation of Ca^{++} channels in different tissues. For example, Cd^{++} and Zn^{++} are both powerful blocking cations for most Ca^{++} channels. However, Cd^{++} is a permeant ion species in muscle fibers of some insect larvae, and a Zn^{++} spike has been observed in some snail ganglia (Kawa 1979). At this point, the molecular basis for these differences is far from being determined.

At least three distinct modes of inactivation of Ca^{++} channels have been seen in different tissues:

1. In some molluscan neurons, there is one type of current-dependent inactivation whereby Ca^{++} influx leads to an increase in [Ca^{++}]$_i$, which blocks the Ca^{++} channel (see Brown; Tillotson; both this volume).
2. In frog skeletal muscle, Palade and Almers (1978) have reported another type of current-dependent inactivation of I_{Ca}, apparently due to depletion of Ca^{++} from the unstirred extracellular spaces.
3. In polychete eggs, Fox (1980) has seen a type of inactivation of the Ca^{++} channel that depends neither on the concentration nor the identity of the divalent ion present and seems to be directly voltage-dependent. This inactivation mechanism is analogous to that described by Hodgkin and Huxley (1952) for the Na$^+$ channels in squid axon (see Brown, this volume).

EXPERIMENTAL DIFFICULTIES IN THE STUDY OF THE Ca^{++} CHANNEL

• There are four problems that arise in biophysical studies of the Ca^{++} channel. First, conditions that permit a good space-clamp during voltage-clamp studies are difficult to obtain in most preparations that have reasonable I_{Ca} (Fig. 1). For this reason, Hagiwara did not attempt a detailed analysis of the time course of I_{Ca} in barnacle muscle fibers. One of the experimental preparations in which space-clamp conditions are most closely approximated is the isolated molluscan nerve cell soma. However, work done with less-favorable preparations, such as barnacle muscle fibers, heart muscle strips, or presynaptic terminals, is not without value. On the contrary, if the limitations on the level of analysis that is possible using these imperfect preparations are clearly understood, then important information can be obtained about the properties of the Ca^{++} channel.

SPACE CLAMP CONDITION

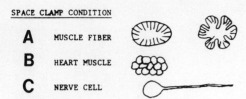

A   MUSCLE FIBER

B   HEART MUSCLE

C   NERVE CELL

Figure 1
Schematic representation of geometric factors that limit the degree to which one can space-clamp cells known to have I_{Ca}s. (A) Somatic muscle fibers have t-tubules or deep infoldings that result in a high series resistance. (B) Heart muscle fiber bundles have narrow clefts between individual fibers that result in a high series resistance value. (C) A nerve cell soma can be space-clamped satisfactorily only if one uses special techniques to eliminate the complications that arise from poor control of V_m in axonal and dendritic processes. Somas isolated by ligaturing or cutting, patch-clamping, and differential electrode-pair recordings all have been used successfully to overcome the problems that arise in clamping soma membrane.

A second problem arises in trying to determine the reversal potential and selectivity of the Ca^{++} channel. Knowledge of the selectivity of either Na^+ or K^+ channels comes from analyzing the effects of different ionic concentrations on the reversal potential of I_{Na} or I_K, respectively. But the I_{Ca} reversal potential, E_{Ca}, is difficult to determine experimentally because the $[Ca^{++}]_i$ is very low—10^{-7} M or less. Therefore, a significant outward I_{Ca} is not expected. Figure 2 shows I_{Ca} as a function of V_m calculated from the constant-field equation (assuming a constant permeability to Ca^{++}). This figure shows how difficult it is to determine E_{Ca} experimentally. One solution to this problem might be to increase the $[Ca^{++}]_i$ so that the Ca^{++} concentration gradient across the membrane becomes smaller (more like the Na^+ concentration gradient across the Na^+ channel), but this does not work because internal Ca^{++} blocks the Ca^{++} channel. This inherent asymmetry in the conduction of Ca^{++} through the Ca^{++} channel is notably different from the behavior of Na^+ in the Na^+ channel, which shows no such asymmetry. This feature is particularly important if one wishes to apply a Hodgkin-Huxley–type analysis to the Ca^{++} channel.

Although the selectivity of the Ca^{++} channel cannot be determined by measuring E_{Ca}, it is possible to learn something about selectivity by measuring the peak currents in solutions of different divalent cations. For solutions with a single divalent ion present, the following sequence is obtained: $I_{Ba} > I_{Sr} > I_{Ca}$. If, however, 20 mM Co^{++} is added to the solutions to block the Ca^{++} channel partially, then the following sequence is obtained: $I_{Ca} > I_{Sr} > I_{Ba}$. This difference can be explained as follows: Suppose that ions must bind to a site within the channel before they can move through it. Then separate binding and mobility factors will affect the ease with which each ionic species passes through the channel. It would be possible to identify these two factors separately only if ions do not move independently through the channel, as is the case for the Ca^{++} channel, which shows saturation. Given these conditions, the presence of a blocking ion such as Co^{++} would be expected to increase the importance of the binding affinity relative to the mobility, thereby changing the selectivity as observed.

The third problem is how to separate I_{Ca} from the counteracting outward current. In their classic studies on squid axon, Hodgkin and Huxley (1952) separated I_{Na} from I_K with the aid of a few simple assumptions. One of their key premises was that altering $[Na^+]_e$ to change I_{Na} affects neither I_K nor the time course of I_{Na}. An analogous assumption can not be made with regard to the Ca^{++} channel because of three complicating factors.

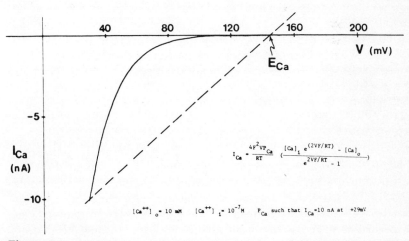

Figure 2
I-V relationship of I_{Ca} calculated from the constant-field equation (———).
The I-V relation that would be expected if Ca^{++} channels had ohmic properties (- - - -). All conditions are listed in the figure.

1. Surface potential may change when $[Ca^{++}]_e$ is altered, resulting in modifications of the gating and rectification properties of the Ca^{++} channels. Therefore, experiments must be done under conditions that minimize changes in surface potential. For example, if the total concentration of divalent cations outside the cell is maintained at a high level by using a medium with a high Mg^{++} content, then changes in $[Ca^{++}]_e$ will cause only minor changes in surface potential.

2. Ca^{++}-induced I_K ($I_{K(Ca)}$) is activated whenever there is an appreciable influx of Ca^{++} (Meech; Lux; both this volume). Therefore, a change in the inward I_{Ca} may alter the outward I_K.

3. Internal Ca^{++} may block the Ca^{++} channel (see Tillotson, this volume). If so, a change in the net Ca^{++} influx may modify the time course of I_{Ca}.

The latter two difficulties may partially be overcome by introducing a high concentration of EGTA into the cell to trap and buffer the incoming Ca^{++}. However, the buffering action of EGTA may not be fast enough to eliminate completely Ca^{++}-induced outward I_K and Ca^{++}-channel inactivation. Therefore, the method adopted most commonly for the isolation of I_{Ca} is to replace intracellular K^+ with impermeant cations such as Cs^+ or $Tris^+$, so as to eliminate the counteracting outward current altogether. This method apparently works if V_m is not made very positive—in other words, if one does not try to obtain E_{Ca}. When V_m is made too positive, a slowly activating outward current appears that is different from $I_{K(Ca)}$.

This slowly activating outward current is the fourth problem. Such a current has been seen in various preparations. It is present in heart muscle, for which Kass and Tsien (1975) showed that any Ca^{++}-channel blocking agent, such as Co^{++}, La^{+++}, or D-600, blocks the slow outward current at the same concentration at which it blocks the Ca^{++} channel. A similar result has been obtained by Palade and Almers (1978) in frog skeletal muscle fibers, where the slow outward current can be carried by Cs^+ or Na^+ as well as by K^+. It might well be asked whether this nonspecific, slowly activating outward current is the so-called $I_{K(Ca)}$ (see Meech; Lux; both this volume). This is unlikely, however, since in both heart and skeletal muscle, an increase in $[Ca^{++}]_e$, which would increase the Ca^{++} influx and, consequently, the amount of internal free Ca^{++}, produces a decrease in the slow outward current. One would expect just the opposite for $I_{K(Ca)}$.

A similar slowly activating outward current has been observed by Hagiwara in internally dialyzed molluscan ganglion cells (Fig. 3). Ganglion cells from the freshwater snail *Lymnaea* were dialyzed internally with Tris-aspartate and EGTA and maintained in a Na^+-free external solution. I_{Ca}s are seen in Figure 3 as slowly inactivating inward currents. When V_m was stepped to large positive values (above $+40$ mV), a time- and voltage-dependent outward current appeared. This slowly activating outward current probably was not carried by Ca^{++}, because the inside of the cell contained EGTA and no Ca^{++}. The presence of internal EGTA also makes it unlikely that this current can be identified with $I_{K(Ca)}$. Further evidence against a contribution of $I_{K(Ca)}$ to this slowly activating outward current comes from experiments with Ba^{++}. Although Ba^{++} is known to be less effective than Ca^{++} for inducing outward I_K, replacing external Ca^{++} with Ba^{++} increased the delayed outward current that was observed. Thus, by exclusion, this current is probably carried by $Tris^+$. In other experiments, Na^+, Cs^+, and even TEA^+ also were able to sustain the outward current.

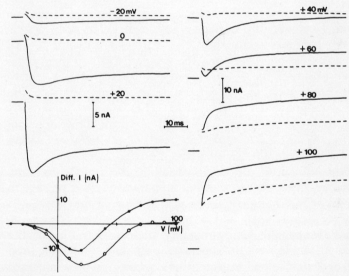

Figure 3
Ca^{++}-dependent, nonspecific outward current. Current records were obtained under voltage-clamp conditions from a *Lymnea* ganglion cell perfused internally with Tris-aspartate and 5 mM EGTA and bathed in Na^+-free saline. Control currents (———); currents recorded from the same cell with 1 mM Cd^{++} added to the external solution to block Ca^{++} channels (----). Note the outward, Cd^{++}-sensitive current at very positive values of V_m. (*Inset*) *I-V* plots showing the time dependence of the Ca^{++}-dependent, nonspecific outward current. Current measured at a point just following the initial rapid change in ionic current (O); currents measured 60 msec after start of pulse (●). (L. Byerly, M. Masuda, and S. Hagiwara, in prep.)

Whereas the slowly activating outward current probably does not depend on a build-up of $[Ca^{++}]_i$, it can be reduced by Ca^{++}-channel blockers. As shown in Figure 3, 1 mM external Cd^{++} not only blocks inward I_{Ca}, it also reduces the delayed outward current at values of V_m above +40 mV. Thus, it is not possible simply to conclude that the difference between the paired traces in Figure 3 is I_{Ca}.

The existence of this voltage- and time-dependent outward current raises two questions: (1) What is this outward current, and how is it related to I_{Ca}? (2) How can I_{Ca} be separated from it? There are a few speculations about the answer to the first question, such as the two currents may go through the same channel. Unfortunately, it is too early to state any conclusions.

SUMMARY

• Many questions about the Ca^{++} channel remain to be answered. As discussed by Tillotson, Brown, and Llinás (all this volume), we know much about the kinetics, mode of inactivation, and selectivity of these channels, but the molecular details of their mechanisms are still to be worked out. The diversity of Ca^{++} channels that exist in different tissues means that these problems may have to be solved more than once. Ultimately, the questions of why the Ca^{++} channel is in so many different tissues, what function it serves, and how it interacts with other parts of the cell need to be answered.

REFERENCES

Fox, A. 1980. Voltage-dependent inactivation of the calcium channel. *Proc. Natl. Acad. Sci.* (in press).

Hagiwara, S. 1973. Calcium spike. *Adv. Biophys.* **4:** 71.

Hagiwara, S. and L. Byerly. 1981. Calcium channel. *Annu. Rev. Neurosci.* (in press).

Hagiwara, S., K. Kusano, and N. Saito. 1961. Membrane changes of *Onchidium* nerve cell in potassium-rich media. *J. Physiol. (Lond.)* **155:** 470.

Hodgkin, A.L. and A.F. Huxley. 1952. The dual effect of membrane potential on sodium conductance in the giant axon of *Loligo. J. Physiol. (Lond.)* **116:** 497.

Kass, O.S. and R.W. Tsien. 1975. Multiple effects of a calcium antagonist on the plateau current in cardiac Purkinje fibers. *J. Gen. Physiol.* **66:** 169.

Kawa, K. 1979. Zinc-dependent action potential in the giant neuron of the snail, *Euhaera quaestia. J. Membr. Biol.* **49:** 325.

Oomura, Y., S. Ozeki, and T. Maena. 1961. Electrical activity of a giant nerve cell under abnormal conditions. *Nature* **191:** 1265.

Palade, G.P. and W. Almers. 1978. Slow sodium and calcium current across the membrane of frog skeletal muscle fiber. *Biophys. J.* **21:** 168A.

Ca^{++}-dependent Inactivation of Ca^{++} Channels

Based on a presentation by

DOUGLAS TILLOTSON

Department of Physiology
Boston University Medical School
Boston, Massachusetts 02118

• The sensitivity of the Ca^{++} channel to elevated [Ca^{++}]$_i$ levels is well known. Perhaps the earliest observation of this phenomenon was that of Hagiwara and Nakajima (1966), who found that when [Ca^{++}]$_i$ levels were elevated to 10^{-5} M, while at the same time the [Ca^{++}]$_e$ was adjusted so as to leave E_{Ca} unchanged, the Ca^{++} action potential in barnacle muscle fiber was abolished. Later, with the development of perfused cell preparations, it was demonstrated directly that raising [Ca^{++}]$_i$ levels into the 100 nM range reduced g_{Ca} (Kostyuk et al. 1977; Akaike et al. 1978). Kostyuk suggested that I_{Ca}s might therefore have a self-blocking action in the intact cell. Here it is proposed that the Ca^{++} influx through Ca^{++} channels opened by depolarization mediates inactivation of those channels. This mechanism of inactivation would be somewhat novel. Although most known voltage-sensitive Na$^+$ and K$^+$ channels show inactivation, in every case the inactivation is considered to be dependent directly on V_m, not on ion fluxes through the channels.

METHODS

• All of the data presented here were obtained from cells R2, R14, and R15 in *Aplysia*. At the beginning of an experiment, the ganglion was exposed to a solution containing Cs$^+$ and nystatin. Nystatin, an ion carrier, created a large permeability to monovalent ions, allowing some of the external Cs$^+$ to exchange for greater than 90% of internal K$^+$. The nystatin was then washed out, leaving Cs$^+$-loaded cells (see

The report of this presentation was prepared by J. Strong.

Tillotson and Horn 1978). Since Cs^+ is an excellent blocker of K^+ channels, this is an effective method for blocking the outward I_Ks that normally would obscure the I_{Ca}s of interest. The cell was then subjected to a standard two-electrode voltage clamp in Na^+-free Ringer's solution containing 100 mM Ca^{++}. Under these conditions, virtually all of the ionic current recorded is carried by Ca^{++}; outward currents were not observed unless the membrane was depolarized to V_ms more positive than +40 mV. Results qualitatively similar to those presented below have been obtained in solutions containing 10 mM Ca^{++}.

Inactivation was studied by the classical two-pulse technique of Hodgkin and Huxley (1952). This method is illustrated in Figure 1. Inward currents were measured during two depolarizations, separated, in this example, by several hundred milliseconds. The peak inward current during the second pulse (PII) is smaller than it would have been had no first pulse (PI) been present. This inactivation is quantitatively defined as the ratio of the PII peak currents seen with and without a first pulse. The smaller the ratio, the greater the degree of inactivation. In the example in Figure 1, the two pulses are identical, so one can observe the degree of inactivation directly by comparing the peak currents seen during PI and PII.

INACTIVATION IS PROPORTIONAL TO Ca^{++} ENTRY

• The basic premise of this discussion is that in a two-pulse experiment, such as the one shown in Figure 1, inactivation (i.e., the reduction of I_{Ca} during the second pulse) is caused by Ca^{++} entry during the first pulse. This conclusion is based on experiments like that illustrated in Figure 2. Here the voltage of PI was varied, and the degree of I_{Ca} inactivation as a function of PI was measured. Figure 2A shows that as the voltage of PI increases from −20 mV to +40 mV, the peak current during PII is reduced progressively. This segment of the inactivation curve resembles the voltage-dependent inactivation

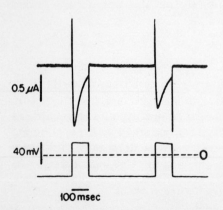

0.5 μA

40 mV

100 msec

Figure 1
I_{Ca} inactivation associated with a pair of identical voltage-clamp pulses in a Cs^+-loaded R2 cell. Peak inward current during PII is reduced by PI. +20-mV, 100-msec pulses separated by 400-msec intervals; $V_h = -40$ mV. (Reprinted, with permission, from Tillotson 1979.)

seen in Na$^+$ and K$^+$ channels. However, as the voltage during PI is increased further, the peak current during PII begins to increase again, until, for PI voltages in the region above 100 mV, no inactivation at all is seen. In this region, the inactivation curve is strikingly different from that seen in most other voltage-dependent channels, which have monotonically decreasing inactivation curves. Also shown in Figure 2A is the Ca^{++} entry during PI at each voltage, which is found by integrating the current observed during PI. With increasing depolarization, Ca^{++} entry first increases, as voltage-dependent activation increases. But with further depolarization, Ca^{++} entry decreases, as the decrease in driving force for Ca^{++} influx begins to predominate. The amount of Ca^{++} entry parallels closely the amount of inactivation seen during PII, as would be predicted by the theory that Ca^{++} entry mediates inactivation. This is shown more clearly in Figure 2B, where the PII current has been plotted directly against Ca^{++} entry. A simple proportionality is found. In another experiment, the amount of Ca^{++} that entered during PI was varied over a 30-fold range by varying the duration of PI (from 5 msec to 200 msec) and the amplitude of PI (from +10 mV to +30 mV). A simple proportionality between Ca^{++} entry during PI and the degree of inactivation also was found under these conditions.

The simplest explanation for the results in Figure 2 is that the Ca^{++} that enters the cell during PI directly causes the inactivation observed during PII. This explanation would be consistent qualitatively with the reduction of I_{Ca} seen in cells perfused with solutions containing elevated [Ca^{++}]. However, two alternative explanations of the results also must be considered.

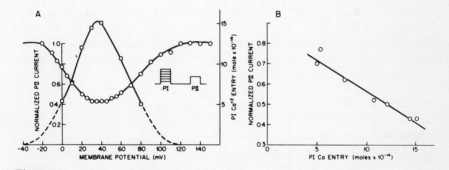

Figure 2
Ca^{++} inactivation related to Ca^{++} entry. V_m of PI is varied; PII V_m is held constant at +20 mV. (A) Normalized PII peak current vs V_m of PI (○). PI Ca^{++} entry (time integral of leakage-corrected PI converted to moles of Ca^{++} entry) (□) vs V_m of PI. (B) PII inactivation as function of extent of Ca^{++} entry during PI. Cs$^+$-loaded R2 cell; V_h = −40 mV. (Reprinted, with permission, from Eckert and Tillotson 1980.)

Driving-force Reductions Cannot Explain the Decrease in I_{Ca}

One explanation of the results seen in Figure 2 is that as Ca^{++} enters the cell and accumulates just inside the membrane, the driving force for Ca^{++}, and hence I_{Ca}, is reduced. Such a reduction in driving force would cause a reduction in I_{Ca} dependent on previous Ca^{++} entry but would not represent an inactivation process intrinsic to the Ca^{++} channels. This explanation of the results seems untenable given that the Ca^{++} channels rectify strongly and do not pass measureable outward currents (see Hagiwara, this volume). Therefore, the size of the net inward I_{Ca} should be relatively independent of the concentration of permeant ions on the inside of the membrane. Experimental evidence that argues against the driving-force explanation is shown in Figure 3. Here inactivation (measured as in Fig. 2) in 100 mM Ba^{++} is compared with that seen in 100 mM Ca^{++}. The currents in Ba^{++} are larger and decay more slowly, so that at a given voltage the Ba^{++} influx during PI exceeds the Ca^{++} influx. A postulated change in driving force would therefore be expected to be much greater for Ba^{++} than for Ca^{++}, yet there is much less inactivation in the Ba^{++} solution. It is unlikely that intracellular buffering of Ba^{++} is so much faster than Ca^{++} buffering that a smaller change in driving force is seen for Ba^{++} in Figure 3, even though the Ba^{++} entry is larger. Arsenazo measurements of $[Ca^{++}]_i$ indicate that the decline of $[Ba^{++}]_i$ following a depolarization is $10-20$ times slower than the decline in $[Ca^{++}]_i$.

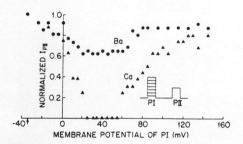

Figure 3
Depression of I_{Ba} is much less than I_{Ca} depression. Pair of 100-msec voltage-clamp pulses separated by 200 msec. V_m of PI is varied; V_m of PII is held constant at +20 mV in 100 mM extracellular Ca^{++} and 100 mM extracellular Ba^{++}. Cs^+-loaded R15 cell; $V_h = -37$ mV. (Reprinted, with permission, from Tillotson 1979.)

Inactivation Represents a True Reduction in I_{Ca}

Another possible explanation of the data presented so far is that the apparent reduction in I_{Ca} seen during PII is actually an increase in an outward current, perhaps in $I_{K(Ca)}$ (see Meech; Lux; both this volume). It is difficult to prove that the blockade of outward current by Cs$^+$ loading is completely effective; in fact, such pharmacological methods are notoriously incomplete. Since $I_{K(Ca)}$ is less strongly activated by Ba^{++} than by Ca^{++}, this explanation of the results cannot be ruled out by the results presented in Figure 3. An experiment to show that the apparent inactivation of I_{Ca} does, in fact, represent a true reduction in Ca^{++} entry is shown in Figure 4. Here the same voltage-clamp protocol as in Figure 2 has been used, but the Ca^{++} indicator arsenazo III (see S. Smith, this volume) has been employed to measure Ca^{++} entry during PI and the degree of inactivation of Ca^{++} influx during PII. The arsenazo signals provide a direct method for measuring Ca^{++}

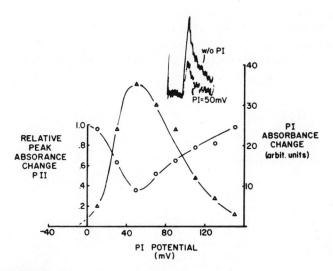

Figure 4
Two-pulse Ca^{++} inactivation measured with arsenazo III. V_m of PI is varied; V_m of PII is held constant at +20 mV; 200-msec pulses separated by 400 msec. The relative peak-PII arsenazo-III signal was determined by subtracting digitally recorded signals with and without PI at each PI V_m and dividing this peak by the peak arsenazo-III signal without PI. This ratio was plotted (O) against PI V_m. Plot of peak-PI arsenazo-III signal against PI potential (Δ). (Inset) PII signal with and without PI of +50 mV. Cell R15 in 100 mM Ca-TEA (50 mM TEA substituted for 50 mM NaCl). $V_h = -40$ mV.

entry, one which makes no assumptions about blockage of outward currents. Nevertheless, the results obtained with arsenazo III are quite similar to those shown in Figure 2: Inactivation of I_{Ca} during PII is proportional to Ca^{++} entry in PI. This experiment also shows that Ca^{++} inactivation occurs in a cell that has not been Cs^+-loaded. An assumption in this experiment is that the $[Ca^{++}]_i$ is always below levels required to saturate the dye.

DECAY OF INACTIVATION SHOWS TWO TIME CONSTANTS

• Presumably, if the time interval between PI and PII is long enough, all of the Ca^{++} that entered during PI will be buffered, sequestered, and (or) removed, and no inactivation will be seen during PII. The time course of the removal of inactivation is shown in Figure 5. Inactivation requires more than 10 sec to decay completely. The data points have been fit with the sum of two exponentials, with time constants of 420 msec and 4.4 sec. These time constants are quite similar to the two fastest time constants reported by Gorman and Thomas (1978), who used arsenazo III to study the decay in $[Ca^{++}]_i$ following a similar depolarization-induced Ca^{++} influx in the same preparation. The similarity of the two sets of time constants provides

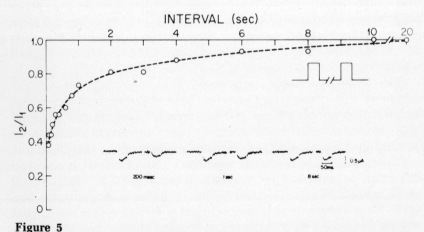

Figure 5
Time course of removal of Ca^{++} inactivation. Pair of identical 50-msec pulses separated by variable intervals. Relative PII current is plotted against the interpulse interval. Sum of two exponential functions with time constants of 420 msec and 4.4 sec (-----). (Inset) Records at three intervals. (Reprinted, with permission, from Tillotson and Horn 1978.)

further support for the idea that [Ca^{++}]$_i$ is the crucial variable governing inactivation of the Ca^{++} channels.

SUMMARY AND CONCLUSIONS

• There are three types of evidence supporting the hypothesis that inactivation of I_{Ca}, as observed during a two-pulse experiment, depends on Ca^{++} entry:

1. Inactivation is proportional to Ca^{++} entry; voltage pulses that vary in amplitude and duration but that result in the same Ca^{++} influx cause the same reduction in peak I_{Ca} measured at a fixed time later.
2. The reduction in I_{Ca} following a Ca^{++} influx is unlikely to be due simply to a change in the driving force for Ca^{++}, since inactivation is much less prominent in Ba^{++} solutions, in which one would expect to see even larger changes in driving force.
3. The observed reduction in I_{Ca} following a Ca^{++} influx is not actually an increase in an outward current, since the reduction in Ca^{++} entry can be observed directly with arsenazo III.

One could postulate that inactivation of Ca^{++} channels, like inactivation of Na$^+$ and K$^+$ channels, is actually a voltage-dependent phenomenon and that it only coincidentally has the same voltage dependence as Ca^{++} entry. This hypothesis seems quite ad hoc, given that a variety of voltage-pulse regimes can be used to cause the same amount of inactivation, provided only that they all cause the same Ca^{++} influx. However, the data are not sufficient to rule out a small voltage-dependent component of Ca^{++} inactivation.

The Ca^{++}-dependent inactivation of I_{Ca} described for *Aplysia* neurons in this presentation also has been observed in *Paramecium* (Brehm and Eckert 1978) and in insect muscle (Ashcroft and Stanfield 1980). However, other mechanisms of Ca^{++} inactivation have been observed in other tissues (see Hagiwara; Brown; both this volume).

This presentation has been concerned with inactivation observed with a two-pulse technique. It seems possible that Ca^{++} entry could also be responsible for the decline in I_{Ca} during a single pulse (see Fig. 1). Consistent with this notion is the observation that the inward currents carried by Ba^{++} decline much more slowly than I_{Ca}s, but it would be interesting to test this idea quantitatively. Another unanswered question concerns the mechanism by which elevated [Ca^{++}]$_i$ levels may block the Ca^{++} channel. The techniques of fluctuation analysis and cell perfusion may prove useful for studying this mechanism (see Brown, this volume).

REFERENCES

Adams, D.J. and P.W. Gage. 1979. Characteristics of sodium and calcium conductance changes produced by membrane depolarization in an *Aplysia* neurone. *J. Physiol. (Lond.)* **289:** 143.

Akaike, N., K.S. Lee, and A.M. Brown. 1978. Calcium current of *Helix* neurons. *J. Gen. Physiol.* **71:** 509.

Ashcroft, F.M. and T.R. Stanfield. 1980. Inactivation of calcium current and skeletal muscle fibres of an insect depends on calcium entry. *Proc. Physiol. Soc.* **C:** 25.

Brehm, E. and R. Eckert. 1978. Calcium entry leads to inactivation of the calcium channel in *Paramecium*. *Science* **202:** 1203.

Eckert, R. and D. Tillotson. 1980. Calcium-mediated inactivation of calcium conductance in caesium-loaded neurones of *Aplysia californica*. *J. Physiol. (Lond.)* (in press).

Gorman, A.F.L. and M.V. Thomas. 1978. Changes in intracellular concentration of free calcium ions in a pacemaker neurone, measured with the metallochromic indicator dye arsenazo III. *J. Physiol. (Lond.)* **275:** 357.

Hagiwara, S. and S. Nakajima. 1966. Effects of intracellular calcium ion concentration upon excitability of the muscle fiber membrane of a barnacle. *J. Gen. Physiol.* **49:** 807.

Hodgkin, A.L. and A.F. Huxley. 1952. The dual effect of membrane potential on sodium conductance in the giant axon of *Loligo*. *J. Physiol. (Lond.)* **116:** 497.

Kostyuk, P.J. and O.A. Krishtal. 1977. Effects of calcium and calcium chelators on inward and outward current on the membrane of molluscan neurons. *J. Physiol. (Lond.)* **270:** 569.

Tillotson, D. 1979. Inactivation of calcium conductance depends on entry of Ca ions in molluscan neurons. *Proc. Natl. Acad. Sci.* **76:** 1497.

Tillotson, D. and R. Horn. 1978. Inactivation without facilitation of Ca conductance in caesium-loaded neurones of *Aplysia*. *Nature* **273:** 512.

Properties of Ca^{++} Channels in Somata of Snail Neurons

Based on a presentation by

ARTHUR M. BROWN*

Department of Physiology
University of Texas Medical Branch
Galveston, Texas 77550

• A first step toward an understanding of the molecular function of the Ca^{++} channel is a detailed quantitative analysis of its electrophysiological properties: voltage sensitivity, opening and closing kinetics, relative permeability to various ions, and single-channel conductance. As pointed out by Hagiwara (this volume), in most excitable cells with prominent I_{Ca}s, it is difficult to obtain accurate records of g_{Ca} as a function of V_m and of time. Brown et al. have largely overcome these technical problems by developing a suction-pipette, voltage-clamp technique for studying the Ca^{++} channels of the nerve cell somata of *Helix aspersa*.

A Hodgkin-Huxley–type description of the kinetics, voltage dependence, and ionic sensitivities of the Ca^{++}-channel activation (m) and inactivation (h) gates was obtained by examining macroscopic I_{Ca}s elicited by depolarizing voltage steps. In addition, fluctuation analysis was applied to the voltage-clamp current records to estimate the conductance of single Ca^{++} channels. By combining the macroscopic and microscopic I_{Ca} data, the density of Ca^{++} channels in the soma membrane was calculated.

METHODS FOR STUDYING Ca^{++} CHANNELS

• Brown et al. used large, identified nerve cell bodies dissected from the subesophageal ganglia of *Helix*. The axon of each neuron was ligatured near the cell body, making it possible to space-clamp the soma. A micropipette was inserted into the cell body to record V_m.

---

*Work presented here was done in collaboration with Y. Tsuda and K. Morimota. *The report of this presentation was prepared by J. Strong.*

For voltage-clamping, a fluid-filled suction pipette was applied to the other pole of the soma, isolating a $15-50$-μm-diameter patch of membrane from the bathing solution. A 1-μm-tip platinum-wire electrode within the suction pipette was used to punch a hole in the isolated membrane patch. The break in the membrane is important because it allows one to modify the intracellular ionic contents by flushing the suction pipette with various solutions. The wire protruding through the hole provides a low-resistance pathway for injecting current into the cell. Because there is a low-resistance pathway to the cytoplasm, the cell can be voltage-clamped by simply clamping the potential of the pipette interior while the potential of the bath is kept at ground.

I_{Ca} was isolated during voltage-clamping by blocking other ionic currents. I_{Na} was suppressed by substituting Tris for external Na^+ and by replacing the cytoplasm with a Na^+-free solution. I_K was eliminated by substituting Cs^+ for intra- and extracellular K^+. Complete blockage of K^+ channels was ensured by adding TEA to both the intra- and extracellular solutions and 4-AP to the extracellular medium. Under these conditions, I_{Ca} is inward at V_ms below about +50 mV, above which potential I_m, due to a nonspecific current, becomes outward. After blockage with external Co^{++}, the outward current is reduced. The nonspecific current is voltage-dependent at V_ms above +50 mV, has a rise time of a few milliseconds, and is about 5 nA in amplitude at +100 mV. It can be ignored in the present experiments.

In *Aplysia* neurons, the influx of Ca^{++} is thought to cause inactivation of the Ca^{++} channels, whereas an inward I_{Ba} through the channels has no such effect (Tillotson, this volume). To test whether *Helix* Ca^{++} channels show a similar selectivity of response to these two ionic species, Ba^{++} was substituted for external Ca^{++} in some experiments.

Properties of single Ca^{++} channels were determined by analyzing the fluctuations in I_{Ca} activated by voltage-clamp steps of $5-15$-sec duration. These depolarizing steps were within the voltage range of -30 mV to -10 mV, voltages at which I_{Ca} (or I_{Ba}) inactivates quite slowly. In the initial experiments of Brown et al. (Akaike et al. 1978b), background instrumentation noise was corrected for by subtracting from the complete noise spectrum a spectrum measured with the I_{Ca} blocked—either by hyperpolarization or by the addition of extracellular Ni^{++} or Co^{++}. In more recent experiments, membrane admittance, which is analogous to frequency-dependent conductance, was determined by applying a white-noise voltage signal to the cell during a depolarizing voltage-clamp step. The background instrumentation noise was calculated by multiplying the membrane admittance by the measured voltage noise of the suction pipette. This calculated background noise was then subtracted from the current power-density spectrum.

MACROSCOPIC CURRENT RECORDS

• The onset of I_{Ca} in response to a depolarizing voltage pulse can be fit by an exponential function with a single time constant, τ_m. The τ_m parameter is a bell-shaped function of voltage that peaks at about -15 mV to -20 mV and becomes quite flat at V_ms above $+15$ mV. The value of τ_m is the same for I_{Ca} and I_{Ba}.

The inactivation of either I_{Ca} or I_{Ba} during a sustained depolarizing pulse can be fit by the sum of two exponential functions, one with a short (τ_{h_1}) and the other with a long (τ_{h_2}) time constant. A major difference between the results presented here and those of earlier experiments (Akaike et al. 1978a) is that in these more recent experiments, larger cells, with larger I_ms, have been used. The fast rate of inactivation $(\tau_{h_1}^{-1})$ was about twice as slow in these cells as in the cells studied previously. Both τ_{h_1} and τ_{h_2} are bell-shaped functions of V_m; they peak at about -15 mV and become quite flat when V_m reaches about $+15$ mV. The value of τ_{h_1} is about $10 \times \tau_m$, and τ_{h_2} is about $50 \times \tau_{h_1}$. The τ_{h_1} for I_{Ca} is about one-half τ_{h_1} for I_{Ba}, but τ_{h_2} for I_{Ba} is only slightly greater than τ_{h_2} for I_{Ca}. After adding EGTA intracellularly to buffer internal Ca^{++} activity between 10^{-7} M and 10^{-9} M, τ_hs for I_{Ca} become similar to τ_hs for I_{Ba}. The fact that manipulations that reduce [Ca^{++}]$_i$ (addition of internal EGTA or external Ba^{++}) also slow down the kinetics of inactivation suggests that, as in *Aplysia* neurons, an increase in [Ca^{++}]$_i$ due to Ca^{++} influx may contribute to inactivation of Ca^{++} channels (see Tillotson, this volume).

The steady-state inactivation, variable h_∞, is a U-shaped function of V_m. It has a value of 1 at hyperpolarized potentials, begins to fall at about -45 mV, and rises again at V_ms above $+50$ mV. It is increased by adding EGTA intracellularly to buffer [Ca^{++}]$_i$. The h_∞ value measured for I_{Ba} is greater than that measured for I_{Ca} over a wide range of depolarizing potentials. All of these characteristics of h_∞ are consistent with the hypothesis that a buildup of [Ca^{++}]$_i$ accounts for inactivation of the Ca^{++} channels. Several additional observations indicate, however, that although increases in [Ca^{++}]$_i$ contribute to inactivation, they are not essential for its occurrence. Convergence of h_∞-V_m curves to a value of 1.0 at potentials above E_{Ca} (which is 120 mV in these experiments) is not observed. The relationship between h_∞ and $\int I_{Ca}dt$ is not inverse as predicted by the [Ca^{++}]$_i$ buildup-inactivation hypothesis, and there is an extended region at positive potentials where the relationship is constant. Large changes in [Ba^{++}]$_i$ from 10^{-2} M to 10^{-6} M reduce I_{Ba} amplitude but do not change its time course. They also reduce I_{Ca} inactivation rates. Thus, intracellular Ba^{++} itself does not enhance inactivation and may interfere with Ca^{++} actions in this respect. Since some inactivation is seen at voltages near E_{Ca} and since inactivation of I_{Ca} is altered by intracellular EGTA, it can be concluded that inactivation of I_{Ca} is both Ca^{++}-dependent and potential-dependent. The contribution of each component is variable, possibly as a result of differences in cell sizes and in intracellular Ca^{++} buffering. However, inactivation of I_{Ba} is solely potential-dependent, since it is unaltered by intracellular EGTA and since large changes in $[Ba^{++}]_i$ levels do not alter inactivation.

The I_{Ca}-dependent component of Ca^{++} inactivation could have at least three causes: a change in the Ca^{++} concentration gradient across the membrane, the actual flux of Ca^{++} through the Ca^{++} channel, or a process that is dependent on $[Ca^{++}]_i$ (see Tillotson, this volume). Since at the same value of V_m, inactivation is faster for I_{Ca} than for a much larger amplitude I_{Ba}, it seems unlikely that a change in concentration gradient accounts for the decline in inward current. Nor is inactivation likely to be a direct result of the inward flux of Ca^{++} through the Ca^{++} channel, since no inactivation is produced by pulses of I_{Ca} less than 5 msec in duration. Thus, by exclusion, the Ca^{++}-dependent component of Ca^{++}-channel inactivation appears to result from an increase in $[Ca^{++}]_i$. The internal Ca^{++} accumulation that contributes to inactivation may be quite localized, since Ca^{++}-mediated inactivation persists at levels of $[EGTA]_i$ that are high enough to block $I_{K(Ca)}$.

The steady-state activation variable, m_∞, overlaps considerably with h_∞ when both are plotted as functions of V_m. There is an e-fold change in m_∞ for about a $15-20$-mV change in V_m. From this result, one may calculate from the Boltzmann equation that the equivalent of one or two electronic gating charges moves through the electric field across the membrane during the activation of each channel (Hodgkin and Huxley 1952). A similar relationship is shown by the h_∞-V_m curve.

The time course of the macroscopic current records can be fit by two different kinetic schemes. One is the Hodgkin-Huxley type described above, with g_{Ca} proportional to mh_1h_2 or to $mh_1 + mh_2$. An equally good fit can be obtained by using m^2 rather than m (Kostyuk and Krishtal 1977). The current records also can be fit by using a kinetic scheme with three coupled variables (Akaike et al. 1978a).

MICROSCOPIC CURRENT RECORDS

• Both the Hodgkin-Huxley and the coupled kinetic models of the macroscopic current records predict current-noise spectra with three corner frequencies. For the mh_1h_2 kinetic scheme, these frequencies are inversely proportional to τ_m, τ_{h_1}, and τ_{h_2}. In earlier experiments (Akaike et al. 1978a), single time-constant spectra for I_{Ba} and I_{Ca} with the corner frequencies corresponding to $\tau_{h_1}^{-1}$ were observed. No $\tau_{h_2}^{-1}$ contributions were observed due to large, low-frequency perturbations, possibly caused by mechanical vibrations. Contributions from

$\tau_m{}^{-1}$ were not observed due to substantial current noise at higher frequencies (above 100 Hz), which resulted from suction-pipette current noise flowing through the membrane impedence. In recent experiments, which have improved high-frequency resolution, a high-frequency corner at 150 Hz corresponding to $\tau_m{}^{-1}$ has been observed.

The absence of m and h_2 components of the current-noise spectrum in the earlier experiments did not reduce estimates of the noise variance by more than a factor of two at worst. Therefore, the earlier estimate that the single-channel conductance, γ_{Ca}, has a value of $2-5 \times 10^{-13}$ S remains valid (Akaike et al. 1978b). Llinás (1977) has arrived independently at a similar figure for γ_{Ca} on the basis of combined electrophysiological and morphological data from squid axon terminals.

The calculated value of γ_{Ca} is only one-tenth to one-twentieth the values estimated for Na$^+$ and K$^+$ channels and suggests that g_{Ca} is impeded. This conclusion is supported by measurements of ion permeability as a function of concentration (Neher and Stevens 1977). Hence, P_{Ca} is dependent upon [Ca^{++}]$_i$ and shows saturation at higher levels of [Ca^{++}]$_e$ (Akaike et al. 1978a). Nevertheless, the conductance mechanism is probably a channel for the following reasons. First, the single-channel conductance value indicates ion transport rates of 10^5-10^6 ions/sec, which are too fast for carriers but which are consistent with a channel mechanism. Second, the shapes of the current-noise spectra also resemble the spectral shapes of known channels, such as the gramicidin molecule.

Ca^{++}-channel density can be estimated from the single-channel conductance and from the maximum Ca^{++} conductance measured when all the Ca^{++} channels are open. Using a chord conductance for the instantaneous I-V curves, a value for g_{Ca} of 10 mmho/cm^2 and a membrane-channel density of $20-200/\mu$m^2 can be calculated.

SUMMARY

• The picture of the Ca^{++} channel that emerges from these and earlier studies (Akaike et al. 1978a,b) is that the channel is selectively permeable to Ca^{++}, Ba^{++}, and Sr^{++} and that conduction of these ions is impeded. Gating is voltage-dependent, and the energy requirements of the h gates may be only slightly less than those of the m gates. The resulting overlap of the m and h functions leads to a persistent inward I_{Ca}. I_{Ca} is regulated, in part, by local accumulation of internal Ca^{++}, which at levels of $10^{-6}-10^{-7}$ M turns off the Ca^{++} channel. Since outward I_{Ca} has never been observed, Ca^{++}-mediated inactivation may also prevent outward I_{Ca} from occurring. There may be two or more sets of Ca^{++} channels, or the channel may have more than one open state as well as several inactivated states.

REFERENCES

Akaike, N., K.S. Lee, and A.M. Brown. 1978a. The calcium current of *Helix* neuron. *J. Gen. Physiol.* **71:** 509.

Akaike, N., H.M. Fishman, K.S. Lee, L.E. Moore, and A.M. Brown. 1978b. The units of calcium conductance in *Helix* neuron. *Nature* **274:** 379.

Hodgkin, A.L. and A.F. Huxley. 1952. A quantitative description of membrane current and its application to conduction and excitation in nerves. *J. Physiol. (Lond.)* **117:** 500.

Kostyuk, P.G. and O.A. Krishtal. 1977. Effects of calcium and calcium-chelating agents on inward and outward currents in the membrane of molluscan neurons. *J. Physiol. (Lond.)* **270:** 569.

Llinás, R.R. 1977. Calcium and transmitter release in squid synapse. *Soc. Neurosci. Symp.* **2:** 139.

Neher, E. and C.F. Stevens. 1977. Conductance fluctuations and ionic pores in membranes. *Annu. Rev. Biophys. Bioeng.* **6:** 345.

A Model of Presynaptic Ca^{++} Current and Its Role in Transmitter Release

Based on a presentation by

RUDOLFO LLINÁS*

Department of Physiology & Biophysics
New York University Medical Center
New York, New York 10016

- The Ca^{++} hypothesis of chemical synaptic transmission was originally proposed by Katz and Miledi (1967a; Katz 1969). It states that Ca^{++} influx through g_{Ca} channels in the presynaptic nerve terminal membrane is responsible for coupling the presynaptic action potential to the release process. Many of their critical supporting experiments were done on the squid stellate ganglion (Katz and Miledi 1967b, 1969), in which both the presynaptic and the postsynaptic cells can be impaled with intracellular microelectrodes (Bullock and Hagiwara 1957). This stellate ganglion preparation is also of particular value for more quantitative studies of the presynaptic Ca^{++} channels and their role in the release of transmitter. By inserting two micropipettes into the presynaptic terminal, it is possible to voltage-clamp the presynaptic membrane. Using this approach, Llinás et al. (1976, 1981a) have obtained sufficient data to produce a detailed mathematical model of the gating and permeation properties of the Ca^{++} channels as well as of the dynamics of the coupling of Ca^{++} influx to transmitter release.

I_{CA} AND THE POSTSYNAPTIC RESPONSE ARE SIMILAR IN VOLTAGE DEPENDENCE AND TIME COURSE

- The presynaptic I_{Ca} and its role in triggering the release process were studied by voltage-clamping the presynaptic terminal while recording the V_m of the postsynaptic cell. TTX and 3-AP were applied

---

*Work presented here was done in collaboration with I.Z. Steinberg and K. Walton. The report of this presentation was prepared by S. Mackey and D. Harris.

to block g_{Na} and g_K, respectively. Under these conditions, a depolarizing voltage-clamp pulse in the presynaptic cell elicits an inward ionic current that is slow compared to g_{Na} (Fig. 1). This inward current is carried by Ca^{++}, as shown by the fact that it is blocked by Cd^{++} (1 mM) or Mn^{++} (50 mM). Associated with this presynaptic I_{Ca} is a postsynaptic depolarization, which provides a measure of transmitter release.

Figure 1A–C shows the time-dependent I_{Ca} and the accompanying transmitter release at several levels of depolarization. For small depolarizations, I_{Ca} increases rather linearly with time, following a brief delay. For larger depolarizing steps, I_{Ca} increases with a sigmoidal time course, eventually reaching a plateau that is maintained for up to 30 msec with no sign of inactivation. The postsynaptic response also has a latency. During the plateau phase of I_{Ca}, the postsynaptic V_m depolarizes at a nearly linear rate. In addition to this on-response that occurs during the voltage pulse, the postsynaptic membrane also shows an off-response at the end of the voltage-clamp pulse. This is especially clear for pulses greater than 60 mV. The transmitter release responsible for the off-response is produced by the Ca^{++} tail current (I_t) that flows at the end of the pulse.

The Ca^{++} currents measured during and following each voltage step show different voltage dependencies. As the depolarizing steps increase in size, the plateau values of I_{Ca} increase steadily for steps of up to 60 mV, indicating a steep voltage dependence for g_{Ca} (Fig. 1A, B). When the voltage steps are increased beyond 60 mV, the plateau values of I_{Ca} decrease as the driving force for Ca^{++} becomes progressively smaller. Finally, at depolarizations of 120–140 mV, the presynaptic I_{Ca} measured during the pulse approaches zero (Fig. 1C). I_ts continue to increase for voltage steps beyond 60 mV (Fig. 1A–C). These I_ts are each the product of a driving force, which is constant, and a conductance factor that depends on how many channels are opened during the depolarizing steps. Therefore, the gradual leveling off of I_ts for V_m steps in the range of 80–110 mV indicates that g_{Ca} saturates at large depolarizations.

The postsynaptic on- and off-responses generally parallel the changes in I_{Ca} with V_m (Fig. 1A–C). The peak of the on-response increases for depolarizations up to 60 mV, then decreases, and finally goes to zero at steps of 120–140 mV, the "suppression potential" (Katz 1969). At levels of V_m where there is an appreciable steady I_{Ca}, the amplitude of the postsynaptic on-response decreases during the pulse (Fig. 1B), presumably as a result of vesicle depletion. The off-response, like the Ca^{++} I_t, continues to increase with increasing depolarization, until it reaches a plateau at voltage steps of 80–110 mV (Fig. 1C). Both the on- and off-responses of the postsynaptic membrane are approximately linear functions of the amplitude of the presynaptic I_{Ca}.

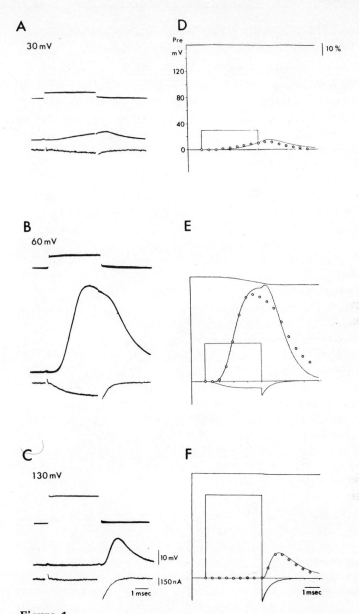

Figure 1
Synaptic transmission during voltage clamp of presynaptic
terminal. (*A,B,C*) Experimental data. (*Top trace*) Presynaptic
voltage; (*middle trace*) postsynaptic response; (*lower trace*)
I_{Ca}. The S-shape of the current onset can be seen at 60-mV
depolarization. Note the fast I_t and the on- and off-EPSP.
(*D,E,F*) Numerical solution to mathematical model. (*Top
trace*) Vesicle depletion; other traces as in *A*–*C*. Recorded
EPSP (○); theoretical curves (———). $[Ca^{++}]_e = 10$ mм. (Reprinted, with permission, from Llinás et al. 1980.)

THE DELAY BETWEEN I_{CA} AND THE POSTSYNAPTIC RESPONSE SETS A LIMIT ON THE EXTENT OF Ca^{++} DIFFUSION

• Voltage-clamping the presynaptic terminal allows one to examine in detail the synaptic delay (the interval between depolarization of the presynaptic terminal and the postsynaptic response). This delay may be divided into two components (Llinás 1977): (1) that for the onset of I_{Ca} that reflects the time required to open the Ca^{++} channels and (2) that between Ca^{++} influx and the onset of the postsynaptic response (Fig. 2). A 60-mV step elicits a postsynaptic on-response that follows the onset of presynaptic depolarization with a latency of 1 msec (Fig. 2A). This latency consists of both of the components of delay. To isolate one of these two components, a depolarizing step was given to the suppression potential. Figure 2B shows the postsynaptic off-response that follows such a step. Since, in this case, the Ca^{++} channels open fully during the depolarizing pulse, the latency between the end of the pulse and the start of the postsynaptic off-response represents only the second component of delay described above and reflects the time required for Ca^{++} to trigger transmitter release. The latency of 200 μsec measured in this way may be a slight overestimate of the second portion of the delay because the falling phase of the voltage pulse is not instantaneous.

The fact that there is such a brief delay (200 μsec) between Ca^{++} influx and transmitter release places an important constraint on the physical picture of the synaptic release process. The distance over which Ca^{++} could diffuse within the terminal in 200 μsec is so limited as to suggest that the site of Ca^{++} influx is quite close to the release site on the presynaptic membrane where vesicle fusion takes place (Llinás and Heuser 1977; Parsegian 1977). The intramembranous particles observed near the vesicle release sites in freeze-fracture studies of the presynaptic terminal have been suggested as possible candidates for the Ca^{++} channels (Dreifuss et al. 1973; D.W. Pumplin et al., in prep.). The observation that the delay between Ca^{++} influx and transmitter release is extremely brief is compatible with the hypothesis that these intramembranous particles may be the Ca^{++} channels (Llinás and Heuser 1977; Parsegian 1977; D.W. Pumplin et al., in prep.).

I_{CA} AND THE RELEASE PROCESS CAN BE DESCRIBED BY A SIMPLE MODEL

The Ca^{++} Channel Has Five Gating Subunits

• Following a few simple assumptions, a mathematical model was developed which describes I_{Ca} and the release process (Llinás et al.

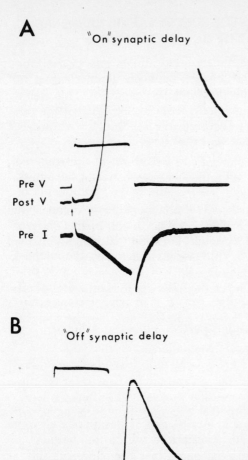

A

"On" synaptic delay

Pre V

Post V

Pre I

B

"Off" synaptic delay

20 mV
5 mV

75 nA

1 msec

Figure 2
Synaptic delay. (*A*) The on-response is shown
for a voltage step of 60 mV from V_R (*upper
trace*). Postsynaptic potential shows a synaptic delay (↑) of 1 msec. The presynaptic
current shows a slow onset and a sharp I_t. (*B*)
The off-delay of about 200 μsec (↑) following
a depolarization to the suppression potential
(140 mV from resting) and the associated I_t.
(Reprinted, with permission, from Llinás
1979.)

1976, 1981b). In this model

$$I_{Ca} = [G] \cdot j \tag{1}$$

where [G] is the number of open channels and j is the I_{Ca} flow through a single open Ca^{++} channel per unit time.

The description of G by this model is based on three assumptions analogous to those adopted by Hodgkin and Huxley (1952) for their model of Na^+ and K^+ channels of squid axon:

1. Each Ca^{++} channel is controlled by n identical, independent, gating subunits. Each subunit alternates between two conformational states—closed (S) and open (S').
2. Depolarization increases, in a noncooperative manner, the probability that each subunit will be in the open conformation, S'. Transitions of a subunit between the open and closed states are governed by voltage-sensitive forward and backward rate constants, k_1 and k_2.
3. All the subunits must be in the open conformation, S', for the channel gate to open.

The system may be described by the following scheme:

$$S \underset{k_2}{\overset{k_1}{\rightleftharpoons}} S'; \; nS' \rightarrow G \tag{2}$$

where k_1 and k_2 are the voltage-dependent forward and backward rate constants for the conversion of S into S' and G designates an open channel. Starting at time 0 with a system in which no channels are open ([G] = 0 and [S'] = 0), the subsequent increase of S' with time is given by

$$[S'] = [S]_0 \frac{k_1}{k_1 + k_2}\left[1 - \exp\{-(k_1 + k_2)\}t\right] \tag{3}$$

where $[S]_0$ is the total number of subunits. The probability that a subunit is in the form S' is given by $[\overline{S}] = [S']/[S]_0$ and that all subunits of a given channel (n) are in the S' form is $[\overline{S}]^n$. The probability of a given channel being open is $[G]/[G]_0$. $[G]_0$ is the total number of channels whether open or closed. Thus,

$$[G]/[G]_0 = \left(\frac{k_1}{k_1 + k_2} [1 - \exp\{-(k_1 + k_2)\}t]\right)^n \tag{4}$$

Combining equations 1 and 4 gives

$$I_{Ca} = [G] \cdot j = [G]_0 \cdot \left(\frac{k_1}{k_1 + k_2} [1 - \exp\{-(k_1 + k_2)\}t]\right)^n \cdot j \tag{5}$$

In the complete form of the equation used in modeling the data, the factor j is expressed as a function of V_m, $[Ca^{++}]_i$, and $[Ca^{++}]_e$.

The rate constants k_1 and k_2 are assumed to vary with V_m, which would mean that the charge distribution of each gating subunit in both the S and S' states differs from its charge distribution in the activation transition state. If it is further assumed that these changes in charge distribution result either from one charged gating molecule per subunit moving through the full electric field of the membrane or from the rotation of a molecule with a dipole moment, then the contributions of V_m to the energies of activation corresponding to k_1 and k_2 are given by

$$z_1 \epsilon V_m = (p - p_1)V_m/m \qquad (6)$$

and

$$z_2 \epsilon V_m = (p - p_2)V_m/m \qquad (7)$$

respectively, where ϵ is the elementary unit of electric charge, m is the membrane thickness, and p, p_1, and p_2 are the components of the dipole moments of a gating subunit in the activated transition state, and the S and S' states, respectively. z_1 and z_2 designate the valences of the hypothetical gating molecules that must cross the membrane during the transformation of each gating subunit from the S or S' state to the activated state. The voltage dependencies of k_1 and k_2 are then given by

$$k_1 = k_1° \exp(\epsilon z_1 V_m/RT); \quad k_2 = k_2° \exp(\epsilon z_2 V_m/RT) \qquad (8)$$

where $k_1°$ and $k_2°$ are constants that do not depend on potential, R is the Boltzmann constant, and T is absolute temperature.

The parameters determining the shape of the I_{Ca} curve are n, $k_1(k_1°, z_1)$, and $k_2(k_2°$ and $z_2)$. Families of curves calculated by computer for different values of these parameters were compared to the experimental results. The best fit was obtained for $n = 5$, $z_1 = 1.42$, $z_2 = -0.38$, $k_1° = 2.76$ msec^{-1}, and $k_2° = 0.14$ msec^{-1}. I_{Ca} curves computed from equation 5 using these values are shown in Figure 1D–F and agree quite well with the experimental data shown in Figure 1A–C. The fact that the best fit of the model was obtained for $n = 5$ suggests that each Ca^{++} channel has five gating subunits.

Transmitter Release during the Action Potential Is an Off-Response

Having modeled I_{Ca}, Llinás et al. next attempted to relate the changes in $[Ca^{++}]_i$ to subsequent transmitter release. The relation between Ca^{++} entry and release was assumed to be mediated by a protein that binds Ca^{++}, causing vesicles to fuse with the membrane and to release their contents into the synaptic cleft. According to the model, Ca^{++} reacts with a fusion promoting factor (fpf) and forms a complex:

fpf · Ca⁺⁺. The concentration of this complex is proportional to $[Ca^{++}]_i$ and the time of exposure to Ca^{++}. The fpf · Ca^{++} complex, in turn, changes by first-order kinetics to an active form, fpf* · Ca^{++}. This active form then triggers vesicle fusion.

Both fpf · Ca^{++} and fpf* · Ca^{++} can revert to inactive states by first-order kinetics. The model is represented by the following scheme:

$$\text{fpf} \cdot Ca^{++} \xrightarrow{k_a} \text{fpf*} \cdot Ca^{++} \xrightarrow{k_i} \text{inactive state} \qquad (9)$$
$$\downarrow k_b$$
$$\text{inactive state}$$

where k_a is the rate constant for activation, and k_i and k_b are the rate constants for the inactivation of fpf · Ca^{++} and fpf* · Ca^{++}, respectively.

The reaction leading to transmitter release can be expressed as

$$TR = k_f \left(\text{fpf*} \cdot Ca^{++} \right) Q \qquad (10)$$

where TR is the rate of transmitter release, k_f is the rate constant for vesicle fusion, and Q is the fractional number of vesicles available for release. It should be noted that TR is directly dependent on fpf* · Ca^{++}. TR is therefore also directly dependent on I_{Ca}.

Finally, three additional assumptions are required to compute the postsynaptic response to a membrane depolarization:

1. There is a very short interval between transmitter release and diffusion across the synaptic cleft.
2. The rate of opening of the conductance channels in the postsynaptic membrane is proportional to the rate of transmitter release.
3. The postsynaptic receptors are well below saturation.

Taking into account the effect of presynaptic membrane depolarization on I_{Ca}, the reaction of Ca^{++} with fpf, the effect of the activated fpf* · Ca^{++} complex on transmitter release, and the effect of neurotransmitter on the postsynaptic V_m, the model can predict the amplitude and time course of the postsynaptic response at various levels of presynaptic voltage clamp. A comparison of the theoretical curves with experimental results (Fig. 1D−F) shows a good fit. Moreover, the model also predicts I_{Ca} and the resulting EPSPs that are caused by a presynaptic action potential (Fig. 3). The calculated curves predicted that transmitter release is basically an off-response, I_{Ca} occurring during repolarization. Indeed, in a recent set of experiments (Llinás et al. 1979) a prerecorded action potential was used as the command voltage (Fig. 3), and the recorded I_{Ca} was seen to agree with the calculated current. Agreement between the postsynaptic responses is good as well. These results indicate that I_{Ca} develops at a time when the Ca^{++} channels are still open, having not yet had time to respond to repolarization, while the driving force for Ca^{++} influx is maximum.

SUMMARY

• These voltage-clamp studies have investigated the characteristics of Ca^{++} channels in nerve terminals, as well as the processes by which Ca^{++} influx is coupled to transmitter release. Two major differences were observed between the Ca^{++} channels in squid nerve terminals and Ca^{++} channels described in various gastropods. First, the squid channels show no appreciable inactivation (see Tillotson, this volume). Second, the squid Ca^{++} channel appears to have five activation gating subunits, compared to the one or two found for gastropod neurons (see Brown, this volume). With regard to the role of Ca^{++} influx in triggering vesicle release, two additional insights emerge. First, the brief delay between Ca^{++} influx and vesicle release suggests that the Ca^{++} channels are quite close to the release sites. Second, the modeling studies reveal that transmitter release triggered by the action potential is an off-response.

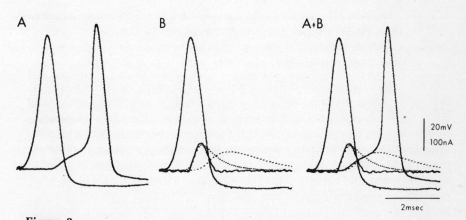

Figure 3
Presynaptic action potential voltage clamp: comparison with model. (A) Experimental pre- and postsynaptic action potentials. (B) Recorded presynaptic action potential used as command voltage, recorded I_{Ca} (——); I_{Ca} and EPSP generated by computer model (-----). (A + B) Comparison of recorded and model-generated I_{Ca} and EPSP (Llinás et al., unpubl.).

REFERENCES

Bullock, T.H. and S. Hagiwara. 1957. Intracellular recording from the giant synapse of the squid. *J. Gen. Physiol.* **40:** 565.

Dreifuss, J.J., K. Akert, C. Sandri, and H. Moor. 1973. Specific arrangements of membrane particles at sites of exo-endocytosis in the freeze-etched neurohypophysis. *Cell Tissue Res.* **165:** 317.

Hodgkin, A.L. and A.F. Huxley. 1952. A quantitative description of membrane current and its application to conduction and excitation in nerve. *J. Physiol. (Lond.)* **117:** 500.

Katz, B. 1969. *The release of neural transmitter substances* (Sherrington Lecture X). C.C. Thomas, Springfield, Illinois.

Katz, B. and R. Miledi. 1967a. Modification of transmitter release by electrical interference with motor nerve endings. *Proc. Roy. Soc. Lond. B* **167:** 1.

————. 1967b. A study of synaptic transmission in the absence of nerve impulses. *J. Physiol. (Lond.)* **192:** 407.

————. 1969. Tetrodotoxin-resistant electric activity in presynaptic terminals. *J. Physiol. (Lond.)* **203:** 459.

Llinás, R.R. 1977. Calcium and transmitter release in squid synapse. *Soc. Neurosci. Symp.* **2:** 139.

————. 1979. The role of calcium in neuronal function. In *The neurosciences: Fourth study program* (ed. F.O. Schmitt and F.G. Worden), p. 555. M.I.T. Press, Cambridge, Massachusetts.

Llinás, R. and J.R. Heuser. 1977. Depolarization-release coupling systems in neurons. *Neurosci. Res. Program Bull.* **15:** 557.

Llinás, R., I.Z. Steinberg, and K. Walton. 1976. Presynaptic calcium currents and their relation to synaptic transmission: Voltage clamp study in squid giant synapse and theoretical model for the calcium gate. *Proc. Natl. Acad. Sci.* **72:** 187.

————. 1980. Transmission in the squid giant synapse: A model based on voltage clamp studies. *J. Physiol. (Paris)* **76:** (in press).

————. 1981a. Presynaptic calcium currents in squid stellate ganglion: A voltage study. *Biophys. J.* (in press).

————. 1981b. Relationship between presynaptic calcium current and postsynaptic potential in squid giant synapse. *Biophys. J.* (in press).

Llinás, R., M. Sugimori, and S. Simon. 1979. Presynaptic calcium current and postsynaptic response generated by a presynaptic action potential; a voltage clamp study in the squid giant synapse. *Biol. Bull.* **157:** 380.

Parsegian, V.A. 1977. Considerations in determining the mode of influence of calcium on vesicle-membrane interaction. *Soc. Neurosci. Symp.* **2:** 161.

Regulation of Intracellular Na⁺, K⁺, and H⁺ Concentrations in Snail Neurons

Based on a presentation by

ROGER C. THOMAS*

Department of Physiology
Yale University School of Medicine
New Haven, Connecticut 06510

● All cells maintain an intracellular environment that is ionically quite different from the external fluid. Maintaining these ionic gradients across the plasma membrane uses a significant portion of the cell's metabolic energy and is quite critical to cell function. In animal cells, for example, the active extrusion of Na^+ is necessary to maintain osmotic equilibrium, the high K^+ levels maintained in the cytoplasm are required by many cell enzymes, and the plasma membrane is rapidly destroyed if $[Ca^{++}]_i$ rises very far above the submicromolar level at which it is maintained normally.

Ionic gradients also play a crucial role in excitability in the neuron. The ionic gradients across neuronal membranes provide the driving forces for the movement of ions through voltage- and transmitter-gated ion channels. Giant molluscan neurons have proved to be excellent preparations in which to study ionic regulation processes. Studies in an identified snail neuron of the Na^+-K^+ pump and of regulation of pH_i are presented here. Regulation of $[Ca^{++}]_i$ is discussed in the presentations that follow (Brinley; S. Smith; both this volume). The interaction between regulation of $[Ca^{++}]_i$ and $[H^+]_i$ is discussed by Meech (this volume).

All the results presented in this chapter were obtained from the soma of an identified cell in the right parietal ganglion of *Helix aspersa* (see Thomas 1977). This cell is large enough (150 μm) to be impaled with up to five microelectrodes at once, permitting simulta-

---

*Current address: Department of Physiology, Bristol University, Bristol BS81TD, England.

The report of this presentation was prepared by J. Strong.

neous voltage measurement, current injection, and measurement of various intracellular ion activities with ion-specific electrodes. Figure 1 shows the concentration gradients found for the major ions in the snail neuron. From these results, values of -81 mV for E_K and $+71$ mV for E_{Na} are predicted. Similar values have been found for marine molluscs, which have much higher values of intra- and extracellular osmolarity.

STOICHIOMETRY OF Na⁺-K⁺ PUMP

• The membrane-bound Na⁺-K⁺ ATPase, which has been found in all animal cells studied so far, has been characterized extensively, mainly through biochemical studies carried out on red blood cells (Glynn and Karlisch 1975). The molluscan neuron has proved to be a useful preparation for studying the pump under more physiological condi-

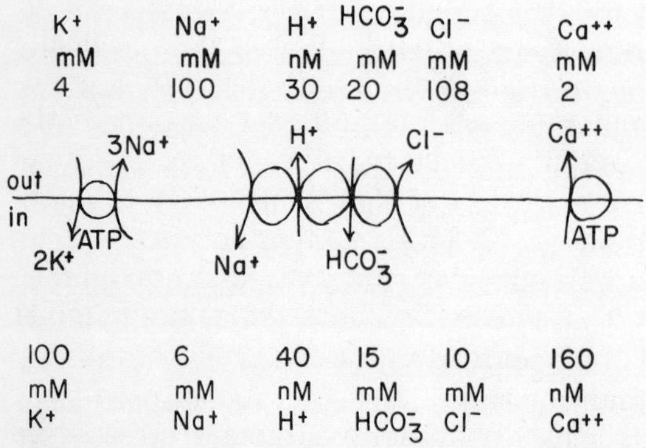

Figure 1
The ionic concentrations in snail hemolymph and neuronal cytoplasm. Arrows show the direction of movement of different ions mediated by the Na⁺-K⁺ ATPase, the H⁺ transport mechanism, and the Ca⁺⁺ pump. Values for snail Ringer's solution are from Thomas (1977). Values for $[K^+]_i$, $[Na^+]_i$, $[Cl^-]_i$, and $[H^+]_i$ were obtained by Thomas (1977 and unpubl.) with ion-specific microelectrodes. The value for $[Ca^{++}]_i$ was obtained by Alvarez-Leefmans et al. (1980) with a newly developed Ca⁺⁺ microelectrode capable of measuring Ca⁺⁺ concentrations in the submicromolar range. $[HCO_3^-]_i$ was calculated from the equilibrium constant for the reaction $HCO_3^- \rightleftharpoons OH^- + CO_2$, by assuming that CO_2 is in equilibrium across the cell membrane.

tions and, in particular, for studying its electrogenic properties. These results will be reviewed briefly here to enable comparison with the pH regulation system described below. All experiments were done under conditions in which the pump should be operating in its normal direction, i.e., moving Na$^+$ out of the cell and K$^+$ into the cell, and using up ATP. That the Na$^+$ pump generates a net charge movement across the membrane is indicated by several different experiments:

1. Blocking the pump, either by removing external K$^+$ or by applying the specific inhibitor ouabain, causes a depolarization of the cell membrane.
2. Stimulation of the pump with an electroneutral injection of Na$^+$ causes a hyperpolarization of the membrane; this hyperpolarization is blocked by ouabain or removal of external K$^+$.
3. In a voltage-clamped cell, the current produced by the pump (following stimulation with a Na$^+$ injection) can be measured directly.

Since, under voltage clamp, the activity of the Na$^+$ pump can be measured directly as a current, quantitative studies of the pump are feasible, as long as only modest injections are made. The current produced by the pump rises linearly during a Na$^+$ injection and declines exponentially following the injection. Simultaneous measurement of $[Na^+]_i$ indicates that the pump current is directly proportional to the $[Na^+]_i$. The net amount of charge moved by the pump in such an experiment is directly proportional to the amount of Na$^+$ injected. Approximately 3 Na ions are pumped out of the cell for each net positive charge that is pumped out; this is consistent with the stoichiometry of 3 Na$^+$:2 K$^+$ seen in the red blood cell experiments (see Thomas 1969). However, the data are not sufficient to rule out slightly different stoichiometries, such as 4 Na$^+$:3 K$^+$. Recent experiments have confirmed these early conclusions, and suggestions that the coupling ratio might be variable have not withstood further investigations (Kononenko and Kostyuk 1976).

REGULATION OF $[H^+]_i$

Regulation of pH$_i$ Is a Multi-ion Process

• Since many enzymes function only in a narrow pH range, and since metabolic processes are constantly producing acid, it might be expected that all cells have some mechanism for regulating pH$_i$. Such a mechanism will, under normal conditions, extrude acid. The value for E_H calculated from the data in Figure 1 is −7 mV; since the V_R is normally −50 mV, the cytoplasm is clearly less acidic than it would be if H^+ were in equilibrium across the cell membrane. In the experiments described below, the pH_i regulation system has been examined by using intracellular pH microelectrodes to study the response of the neuron to an acid load under various ionic conditions. pH_i regulation is found to be somewhat more complex than the Na^+-K^+ pump system described above in that at least three different ions are required for normal recovery from an acid load.

External Na^+

The experiment illustrated in Figure 2 shows that pH_i regulation requires the presence of external Na^+ but not K^+. The cell has been impaled with two KCl microelectrodes, an HCl electrode, and a pH-sensitive electrode. When current is passed from the HCl electrode to the KCl electrode, a net electroneutral injection of HCl occurs. This causes a transient decrease in pH_i, from which the cell recovers. This recovery process is virtually unaffected when K^+ is removed from the Ringer's solution. (The transient hyperpolarization seen when K^+ is returned to the bathing solution is due to the reactivation of the Na^+ pump.) However, when external Na^+ is replaced by Li^+, the recovery from the injection of acid is slowed markedly. The same result was obtained in another experiment (not shown) in which an organic Na^+ substitute, BDAC, was used. It seems likely that external Na^+ is not

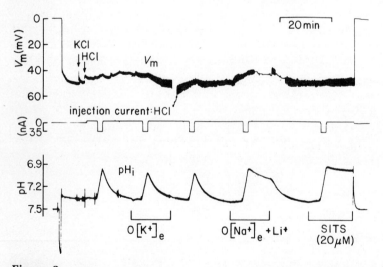

Figure 2
Pen-recordings of V_m injection current, and pH_i from an experiment showing the effect of K^+ removal, Na^+ removal, and SITS on pH_i recovery from HCl injection (Thomas 1976). Ringer's solution buffered with CO_2:HCO_3 (pH 7.5 throughout). Impalement of cell with KCl and HCl electrodes ($\downarrow$).

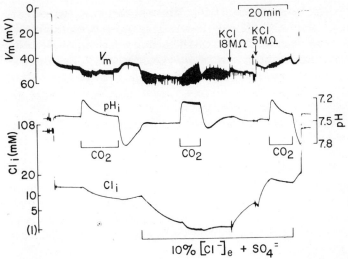

Figure 3

Recording of V_m, pH_i, and internal $[Cl^-]_i$ during experiment showing the effect of reducing $[Cl^-]_i$ on pH_i recovery from CO_2 application. During the period indicated, the estimated $[Cl^-]_i$ was reduced to 10% by replacement of 90% external Cl^- with $SO_4^=$. The low end of the $[Cl^-]_i$ scale is unreliable (Thomas 1977).

simply a cofactor required by the system, but that it is actually transported into the cell as acid is extruded. An experiment similar to that shown in Figure 2, in which $[Na^+]_i$ was also measured, showed that $[Na^+]_i$ rose transiently during recovery from an HCl injection.

Figure 2 also shows that pH_i regulation can be blocked even more effectively with SITS, a well-known blocker of anion exchange in red blood cells. This result indicates that, in addition to Na^+ exchange, H^+ regulation may also involve the exchange of anions across the cell membrane.

$[Cl^-]_i$ and External HCO_3^-

In Figure 3, $[Cl^-]_i$ has been measured with a Cl^--sensitive microelectrode during an acid load. In this instance, an increase in the CO_2 pressure has been used to shift the equilibrium of the $HCO_3^- \rightleftharpoons CO_2 + OH^-$ reaction and acidify the cell. The cell recovers from this acidification quite rapidly and shows a rebound alkalinization when the CO_2 is removed. During the first acid load, a fall in $[Cl^-]_i$ is seen during the recovery from acidification. $[Cl^-]_i$ was then lowered by removing external Cl^-, and the rate of pH_i recovery during CO_2 exposure was reduced considerably. Figure 3 shows that this effect was reversed when $[Cl^-]_i$ levels were restored through impalement with a leaky KCl electrode. These results indicate that normal pH_i regulation requires internal Cl^- and that Cl^- is removed from the cell during recovery from acidification. Removal of external HCO_3^- also blocked the pH_i recovery from a CO_2-induced acid load.

The results presented so far are shown schematically in Figure 4. The dependence of the proposed carrier on $[Na^+]_e$, $[HCO_3^-]_e$, and $[Cl^-]_i$ has been demonstrated above. Since transport of HCO_3^- into the cell is equivalent to moving OH^- into the cell, the system would be capable of removing acid from the cell even if no H^+ were cotransported. In fact, it would be quite difficult to prove or disprove that H^+ itself was transported physically by the pH_i regulatory system. H^+ has been included in the model as a simple way to make the system electroneutral. The system is thought to be electroneutral because pH_i recovery occurs with no change in V_m, and changes in the latter do not affect recovery rate (Thomas 1978).

pH_i Regulation Does Not Require ATP

It is difficult to obtain evidence to show that the simple stoichiometry indicated in Figure 4 (1 $HCO_{3\,e}^-$:1 Na_e^+:1 Cl_i^-:1 H_i^+) is correct. The existence of pH buffers in the cytoplasm makes it difficult to correlate observed pH_i changes with the amount of acid actually extruded, as would be necessary to determine the stoichiometry of the system. However, the carrier with the stoichiometry proposed in Figure 4 would be thermodynamically capable of operating without additional energy sources, such as ATP. The energy derived from moving Na^+ down its electrochemical gradient would be more than adequate to move equimolar amounts of H^+, Cl^-, and HCO_3^- against their electrochemical gradients. The experiment shown in Figure 5 demonstrates that vanadate, a potent inhibitor of many phosphatases, including the Na^+ pump and the Ca^{++} ATPase (Macara 1980), has no effect on the

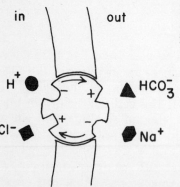

Figure 4
Snail neuron pH_i regulation; proposed four-ion carrier.

ability of the cell to recover from an acid load. The vanadate injection was sufficient to block the Na⁺ pump, however, as shown by the rise in [Na⁺]ᵢ following the injection. In other experiments, it was found that a 15-min exposure to the metabolic inhibitor CCmP (carbonyl cyanide m-chlorophenyl hydrazone) (10 μM) did not affect the ability of the cell to recover from an HCl injection, although it clearly blocked the Na⁺ pump (Thomas 1978). CCmP, which inhibits mito-chondria by increasing their permeability to H⁺, caused a slight acidification of the cell; however, normal rates of recovery back to this new pHᵢ level were seen following HCl injections. These two results suggest that the pHᵢ regulatory system described above does not directly require metabolic energy in the form of ATP.

SUMMARY

● Regulation of pHᵢ is a complex process in snail neurons. At least three different ions are required so that normal recovery from acidifi-cation can occur: external Na⁺ and HCO₃⁻, and internal Cl⁻. The stoichiometry of this electroneutral pumping process has not been determined. The recovery process does not seem to require ATP, since it is quite insensitive to vanadate and CCmP. A four-ion carrier with simple stoichiometry can explain the observations and would be

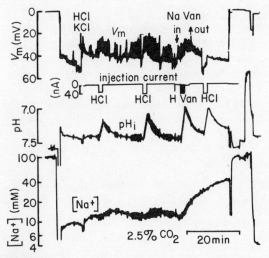

Figure 5
Recording of V_m injection current, pHᵢ, and [Na⁺]ᵢ to show the effect of vanadate injection (by leakage from a microelectrode) on pHᵢ recovery from H⁺ injection (R.C. Thomas, unpubl.).

capable of using the energy of the Na^+ gradient to move other ions against their electrochemical gradients. However, the four-ion model is not a unique explanation of the data, and the model could be tested further by investigating the stoichiometry of the system.

The complexity of the pH_i regulation system is of practical interest to electrophysiologists. Na^+- and HCO_3-free media, which are commonly used, will clearly block pH_i regulation. Since many enzymes are affected critically by pH, processes studied in such media may not be operating in their normal physiological fashion. Conversely, recovery from acidification in cells that have an intact regulatory system may have unforeseen effects: for example, a fall in $[Cl^-]_i$, which will alter the E_{rev} for Cl^--mediated synaptic responses; and a rise in $[Na^+]_i$, leading to stimulation of the Na^+ pump with subsequent changes in E_K and V_m.

REFERENCES

Alvarez-Leefmans, S.J., T.J. Rink, and R.Y. Tsien. 1980. Intracellular free calcium in Helix aspersa neurones. J. Physiol. (Lond.) **306:** 19P.

Glynn, I.M. and S.J.D. Karlisch. 1975. The sodium pump. Annu. Rev. Physiol. **37:** 13.

Kononenko, N.I. and P.G. Kostyuk. 1976. Further study of potential dependence of sodium-induced membrane current in snail neurons. J. Physiol. (Lond.) **256:** 601.

Macara, I.G. 1980. Vanadium: An element in search of a role. TIBS **5:** 92.

Thomas, R.C. 1969. Membrane current and intracellular sodium changes in a snail neurone during extrusion of injected sodium. J. Physiol. (Lond.) **201:** 495.

_____. 1976. Comparison of the Na^+ and H^+ pumps in a snail neurone. J. Physiol. (Lond.) **263:** 212P.

_____. 1977. The role of bicarbonate, chloride and sodium ions in the regulation of intracellular pH in snail neurones. J. Physiol. (Lond.) **273:** 317.

_____. 1978. Comparison of the mechanisms controlling intracellular pH and sodium in snail neurones. Resp. Physiol. **33:** 63.

Ca^{++} Buffering in Squid Axoplasm

Based on a presentation by

FLOYD J. BRINLEY

Neurological Disorders Program
National Institute of Neurological and
Communication Disorders and Stroke
Bethesda, Maryland 20205

• Many intracellular enzymatic reactions are quite sensitive to $[Ca^{++}]_i$. If $[Ca^{++}]_i$ rises above its normally low levels, cell death may result. In nerve cells, $[Ca^{++}]_i$ has an added importance because of its role in coupling excitation to transmitter release. Given these considerations, it is not surprising that nerve cells have highly developed cytoplasmic Ca^{++}-buffering systems. The efficiency and subcellular localization of the Ca^{++}-buffering system in the axoplasm of squid (*Loligo peali*) are described here. Features of the Ca^{++}-buffering process in the somata of gastropod neurons are described by S. Smith and Meech (both this volume).

Ca^{++}-BUFFERING CAPACITY

• The squid axon can be loaded with Ca^{++} by incubation in Ca^{++}-choline seawater, and total Ca load can be measured by analytical techniques. Previous work has demonstrated that when a Ca load is imposed on an axon, only a small fraction appears as free Ca^{++} (Brinley et al. 1977). Using the metallochromic indicator arsenazo III as a monitor of $[Ca^{++}]_i$, the rise in steady-state $[Ca^{++}]_i$ was found to be approximately 1 nM/2 μM internal load. In other words, approximately 99.95% of the imposed load was buffered. As can be seen in Figure 1, $[Ca^{++}]_i$ increased linearly with total Ca loads approaching 1 mmole/kg (20 times the endogenous Ca content of squid axons) without any indication of saturation of the Ca^{++}-buffering system. Thus, squid axons display a large capacity for buffering Ca^{++}.

The report of this presentation was prepared by S. Mackey.

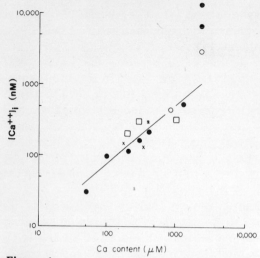

Figure 1
Collected data showing relation between $[Ca^{++}]_i$ and total Ca content. Axons were loaded with Ca^{++} either by stimulation in high-Ca^{++} seawater or by soaking in Na^+-free solutions. (x) Aequorin; ($\bullet$) arsenazo III (A3); (o) A3, apyrase; ($\square$) A3, in vitro; ($\star$) BHR. (Reprinted, with permission, from Brinley et al. 1977.)

In another experiment, the changes in total Ca content and in $[Ca^{++}]_i$ were measured as functions of $[Ca^{++}]_e$. When axons were placed in 3 mM Ca^{++} seawater, the total Ca content rose slowly over several hours from 50–75 μM to 100 μM total Ca. Bathing axons in 10 mM Ca^{++} caused a more rapid and dramatic increase in total Ca content. Changes in the glow of aequorin were used to monitor changes in $[Ca^{++}]_i$. Using different $[Ca^{++}]_e$ levels, it was found that at 3 mM $[Ca^{++}]_e$ there was no change in glow. Such a result suggests that at 3 mM $[Ca^{++}]_e$, squid axon is able to maintain a constant level of $[Ca^{++}]_i$. Ca^{++} concentration in squid hemolymph generally has been reported to be approximately 10 mM. However, Blaustein (1974) has shown that when this concentration is corrected for the presence of sulfate and phosphate, the actual Ca^{++} concentration is 3–4 mM. Taking this into consideration, the results presented here show that the axon is able to maintain constant levels of $[Ca^{++}]_i$ at the in vivo $[Ca^{++}]$ found in squid hemolymph.

SITES OF CA^{++} BUFFERS

• There are four likely sites of Ca^{++} binding in squid axon: (1) small organic anions (aspartate, glutamate, phosphate, citrate, and ATP), (2) soluble proteins, (3) mitochondria, and (4) a CN- and FCCP (carbonyl cyanide p-trifluoromethoxy-phenyl hydrazone)-insensitive buffer system, thought to be the endoplasmic reticulum (ER) (Fig. 2).

As shown in Figure 2, neither organic anions nor soluble proteins are quantitatively significant buffers of Ca^{++}. In regard to organic anions, the Ca^{++}-binding dissociation constants (K_Ds) of aspartate and glutamate (both 500 mM) are too high to be effective physiological buffers; in addition, the concentrations of ATP, citrate, and phosphate are too low to buffer significant quantities of Ca^{++}. In aggregate, these ions could complex only a few hundred nmoles of Ca^{++}/kg axoplasm, or less than 1% of the Ca content in the presence of 30 nM $[Ca^{++}]_i$.

Similarly, the Ca^{++}-binding protein isolated by Alema et al. (1973) from the axoplasm of *Loligo vulgaris* probably can be excluded on quantitative grounds as a significant component of the buffering system. This protein has two binding sites under physiological conditions, each with a K_D of 25 μM. It could therefore bind only about 70 nmoles of Ca^{++}/kg axoplasm at physiological levels of $[Ca^{++}]_i$, and not more than 360 μmoles at saturation. Since the axon can buffer at least 2250 μmoles of Ca^{++}/kg axoplasm without any signs of saturation, the capacity of the binding protein would be inadequate to buffer large loads and its K_D would be too high to take up significant amounts of Ca^{++} at physiological levels of $[Ca^{++}]_i$.

Baker and Schlaepfer (1975) have reported another Ca^{++}-binding

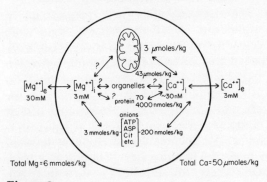

Figure 2
Schematic diagram showing some relations between intracellular and extracellular divalent cations in squid axon with no exogenous load. Intramitochondrial Ca^{++} is shown on the supposition that all of the Ca content of in situ axons, except that existing free or bound to CN- and oligomycin-insensitive entities, is dissolved in mitochondrial water.(Modified from Brinley 1978.)

entity with a lower K_D of approximately 0.5 μM. However, its buffering capacity is only 35−50 μmoles/kg axoplasm. This substance
could bind no more than several μmoles/kg of Ca^{++}/kg axoplasm at
physiological levels of $[Ca^{++}]_i$. Taken together, both the Alema protein and the Baker and Schlaepfer entity would bind no more than
10% of the total Ca in fresh axons and would not bind significant
amounts of Ca at high loads. Thus, the major Ca^{++} buffers in squid
axon appear to be the mitochondria and the CN- and FCCP-insensitive buffering system, which in this presentation is identified with
the ER.

MITOCHONDRIAL RATE AND THRESHOLD OF Ca^{++} UPTAKE

• Previous measurements of mitochondrial uptake of Ca^{++} as a function of $[Ca^{++}]_i$ show a typical sigmoidal curve. However, such measurements extend down only to the multimicromolar range of
$[Ca^{++}]_i$ — well above physiological levels. Therefore, Brinley et al.
performed a series of experiments to measure the threshold value of
$[Ca^{++}]_i$ for mitochondrial uptake of Ca^{++} and the rate of uptake at
physiological concentrations. The protocol was as follows: Arsenazo
III was used to measure changes in $[Ca^{++}]_i$. The axon was exposed to
2 mM CN and various levels of $[Ca^{++}]_e$. Brinley et al. (1978) have shown
previously that CN inhibits initial Ca^{++} uptake by mitochondria. After
the fiber was exposed to Ca^{++}-free seawater containing 2 mM CN,
there was an immediate release of endogenous Ca^{++} from the mitochondria, and $[Ca^{++}]_i$ rose to 0.8 μM. The fiber was then exposed to
various concentrations of Ca^{++}-CN seawater until a steady-state level
of $[Ca^{++}]_i$ was reached, indicating equilibration throughout the cytoplasm. The fiber was then exposed to Ca^{++}-free seawater without CN.
The $[Ca^{++}]_i$ decreased abruptly as Ca^{++} was taken up, presumably by
mitochondria. The maximum slope of the $[Ca^{++}]_i$ versus time curve
was taken as the rate of mitochondrial sequestration of Ca^{++} at a
particular uniform level of $[Ca^{++}]_i$. The results of these and other
experiments are shown in Figure 3. The rate of Ca^{++} buffering attributed to mitochondria is nearly zero until the $[Ca^{++}]_i$ rises to about
200−300 nM. At 1−3 μM $[Ca^{++}]_i$, the rate of uptake is approximately 1
μmole Ca^{++}/(kg axoplasm • min). Above this level, there is a sigmoidal
rise to approximately 20−30 μmoles Ca^{++}/(kg axoplasm • min) at 50
μM $[Ca^{++}]_i$ levels. The clear implication of these results is that mitochondria are not important buffers in the physiological range of
$[Ca^{++}]_i$ in squid axons.

These results also suggest an explanation for differences in the
measured release of mitochondrial Ca^{++} determined by Baker et al.
(1971) and DiPolo et al. (1976). Baker et al. previously reported a
significant release of Ca^{++} from mitochondria, as measured by ae-

quorin, after injecting squid axons with FCCP or CN. DiPolo et al., on the other hand, observed a much more modest release using a similar technique. However, Baker et al. had presoaked their squid axons in 10 mM Ca^{++} seawater before they applied CN. At this level of [Ca^{++}]$_e$, the mitochondrial uptake system was activated and the mitochondria were loaded with Ca^{++}. DiPolo et al., however, did not presoak the axons in 10 mM Ca^{++}, and their results reflect the low mitochondrial Ca^{++} content under physiological conditions. These experiments suggest that mitochondria release significant amounts of Ca^{++} only if they have been loaded previously, and that at physiological levels of [Ca^{++}]$_e$, mitochondria contain only a small amount of Ca.

All of these experiments reflect the average mitochondrial uptake of Ca^{++} and do not measure local variations in mitochondrial uptake. Recently, Mullins and Requena (1979) have provided evidence that mitochondrial uptake may play a more significant role in Ca^{++} buffering near the membrane, where Ca^{++} influx causes a high enough local [Ca^{++}]$_i$ to activate the mitochondrial uptake system. They injected aequorin into squid giant axons to measure changes in [Ca^{++}]$_i$. This response was compared with the response of axons that had also been injected with phenol red. Since the absorption spectrum of phenol red is near the emission spectrum of the aequorin-Ca complex, there is a high probability that photons, emitted by aequorin when it reacts with Ca^{++} in the center of the axoplasm, will be absorbed by phenol red before they can escape from the axon. Photons emitted at the periphery, however, are much less likely to be absorbed. Thus, vari-

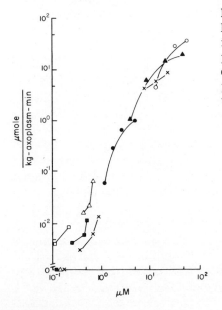

Figure 3
Double logarithmic plot of data showing calculated rate of Ca^{++} uptake by presumed mitochondria as a function of initial ionized Ca. Logarithmic compression makes sigmoidal nature of the curve less obvious than linear plot. (Modified from Brinley et al. 1978.)

ations in the aequorin response measured in the presence of phenol red should reflect changes in [Ca^{++}] near the rim of the membrane. When an axon injected with aequorin and phenol red was stimulated electrically in 50 μM Ca^{++} seawater for 1 min (to allow Ca^{++} entry), there was scarcely a change in [Ca^{++}]$_i$. However, the subsequent addition of 2 mM CN caused a substantial increase in [Ca^{++}]$_i$, suggesting that a large amount of mitochondrial buffering of Ca^{++} occurs near the periphery.

TRANSLOCATION OF Ca^{++} SEQUESTERED BY MITOCHONDRIA

• There is evidence that Ca^{++} initially taken up by mitochondria is transformed to a state that is insensitive to CN. As noted earlier, if an axon is loaded with Ca and then immediately exposed to CN, there is an abrupt rise in [Ca^{++}]$_i$. In another experiment, however, there was a 40-min interval between Ca loading and CN exposure. In contrast to immediate CN application, and in accordance with previous reports, delayed application of CN produces no immediate effect on [Ca^{++}]$_i$. The axon, however, was still sensitive to FCCP, which produced a large increase in [Ca^{++}]$_i$. The simplest conclusion to draw from these experiments is that although the process of sequestering a Ca load as it is being applied requires respiration, the mitochondrial Ca is subsequently changed to a state that is insensitive to CN. In this state, retention of Ca within the mitochondria is still dependent on metabolic energy, as uncoupling by FCCP produces an immediate, substantial release of Ca^{++}. In another series of experiments, in which axons were exposed to CN at various intervals after loading with Ca, it was found that the half time for the process of Ca^{++} translocation to CN-insensitive storage sites within the mitochondria was 3–10 min. Moreover, after longer intervals following loading, the mitochondrial Ca exhibits decreased sensitivity to FCCP treatment. The kinetics of this second translocation have not yet been determined.

CN- AND FCCP-INSENSITIVE BUFFERS

• Results of experiments using FCCP and CN show that a significant amount of buffering of a Ca load occurs in the absence of functioning mitochondria. Thus, it is likely that the ER is this FCCP- and CN-insensitive buffering system. If the mitochondria are inactivated by either FCCP or CN before Ca loading, then subsequent loading will be partitioned between Ca^{++} and the Ca bound to the CN- and FCCP-insensitive system. The results of experiments in which mitochondria were inactivated in this manner showed a nearly linear rise of [Ca^{++}]$_i$ for increasing Ca loads. The slope of the curve indicated that approximately 6% of the load appeared as Ca^{++}; evidently, the remainder had been buffered by the CN- and FCCP-insensitive system. The buffering capacity of the CN- and FCCP-insensitive system is nowhere near saturation even at loads 50 times the normal Ca content.

By measuring the increment in [Ca^{++}]$_i$ produced by FCCP in fresh axons, it is possible to estimate the amount of Ca contained in mitochondria after correcting for buffering by the CN- and FCCP-insensitive system. Results of these calculations indicate that the mitochondrial content of endogenous Ca in fresh axons is about 3 μmoles/kg. Since the total Ca content is approximately 50 μmoles/kg and organic ions and soluble proteins buffer no more than 10% of the Ca load, this suggests that the CN- and FCCP-insensitive system contains approximately 42 μmoles/kg Ca in the fresh axon, or approximately 84% of the total Ca. Results of experiments using axons loaded with Ca show that the CN- and FCCP-insensitive system buffers approximately 60% of the imposed Ca load at values up to millimolar levels. This conclusion is consistent with the results of the experiments on mitochondrial uptake, which suggest that under physiological conditions most of the Ca is not buffered by mitochondria.

In another group of experiments, axons were loaded with Ca and the axoplasm was removed and centrifuged for several hours. About 20% of the initial volume sedimented as a pellet. This included both the mitochondria and the ER, which is presumed to be the CN- and FCCP-insensitive buffer. The remaining 80% of the axoplasm represented the soluble portion of the axon. The fraction of Ca^{++} recovered in the soluble versus organelle compartments was measured as a function of the radiolabeled Ca load. As Ca loading increased, more label was recovered in the pellet. The threshold value of [Ca^{++}]$_i$ for pellet uptake was in the 10-nM range. Thus, in comparison to mitochondria, the CN- and FCCP-insensitive buffer has a lower threshold for Ca^{++} uptake, in addition to its high affinity and large capacity.

As noted previously, the CN- and FCCP-insensitive buffering system has been identified with the ER. Evidence for this hypothesis is based on ultrastructural studies that reveal Ca in what is thought to be the ER. Oschman et al. (1974) have demonstrated the existence of calcium phosphate deposits in ER adjacent to the axolemma of squid axons that had been fixed in glutaraldehyde solution containing 5 mM Ca^{++}. Similar deposits were seen in squid axons by Hillman and Llinás (1974). Finally, Henkart (1975) has described an ER-like structure adjacent to the axolemma which swells under conditions in which the Ca^{++} content of the axoplasm increases.

SUMMARY

• Squid axoplasm has a high buffering capacity at physiological levels of [Ca^{++}]$_i$; it is able to bind 99.95% of a Ca load. The major buffering systems appear to be the mitochondria and the CN- and FCCP-insensitive system, which is identified here with the ER. The mitochondria have a relatively high threshold for Ca^{++} uptake—200–300 nM $[Ca^{++}]_i$—and are not a major Ca^{++} buffer at physiological levels. Despite inhibition with FCCP and CN, the axoplasm is still able to buffer 95% of a Ca load through its CN- and FCCP-insensitive system, which shows a low threshold, high affinity, and large capacity for Ca^{++} binding. There is, however, some evidence that mitochondria near the surface of the membrane may play a more important role in Ca^{++} buffering than do those distributed throughout the rest of the axoplasm.

REFERENCES

Alema, S., P. Calissano, G. Rusca, and A. Giuditta. 1973. Identification of a calcium-binding, brain specific protein in the axoplasm of squid giant axons. *J. Neurochem.* **20:** 681.

Baker, P.F. and W. Schlaepfer. 1975. Calcium uptake by axoplasm extruded from giant axons of *Loligo. J. Physiol.* (*Lond.*). **239:** 37P.

Baker, P.F., A.L. Hodgkin, and E.B. Ridgway. 1971. Depolarization and calcium entry in squid axons. *J. Physiol.* (*Lond.*). **218:** 709.

Blaustein, M.P. 1974. The interrelationship between sodium and calcium fluxes across cell membranes. *Rev. Physiol. Biochem. Exp. Pharmacol.* **70:** 33.

Brinley, F.J., Jr. 1978. Calcium buffering in squid axons. *Annu. Rev. Biophys. Bioeng.* **7:** 363.

Brinley, F.J., Jr., T. Tiffert, and A. Scarpa. 1978. Mitochondria and other calcium buffers of squid axon studied in situ. *J. Gen. Physiol.* **72:** 101.

Brinley, F.J., Jr., T. Tiffert, A. Scarpa, and L.J. Mullins. 1977. Intracellular calcium buffering capacity in isolated squid axons. *J. Gen. Physiol.* **70:** 355.

DiPolo, R.J. Requena, F.J. Brinley, Jr., L.J. Mullins, A. Scarpa, and T. Tiffert. 1976. Ionized calcium concentrations in squid axons. *J. Gen. Physiol.* **67:** 433.

Henkart, M. 1975. The endoplasmic reticulum of neurons as a calcium sequestering and releasing system: Morphological evidence. *Biophys. Abstr.* **15:** 267A.

Hillman, D.E. and R. Llinás. 1974. Calcium-containing electron-dense structures in the axons of the squid giant synapse. *J. Cell Biol.* **61:** 146.

Mullins, L.J. and J. Requena. 1979. Calcium measurement in the periphery of axon. *J. Gen. Physiol.* **74:** 393.

Oschman, J.L., T.A. Hall, P.D. Peters, and B.J. Wall. 1974. Association of calcium with membranes of squid giant axon. *J. Cell Biol.* **61:** 156.

Ca^{++} Regulation in Gastropod Nerve Cell Bodies

Based on a presentation by

STEPHEN SMITH

Department of Physiology and Anatomy
University of California
Berkeley, California 94720

• The regulation of Ca^{++} levels within the cell bodies of gastropod neurons is of particular interest for four reasons:

1. The cell bodies of many gastropod neurons have been shown to have relatively high values of membrane g_{Ca}—apparently much higher than in their axonal membranes. Even a brief train of action potentials in such a cell can produce a sizable influx of Ca^{++} into the soma. If unbuffered, this influx could raise the [Ca^{++}]$_i$ level by several orders of magnitude.
2. The nucleus and its associated synthetic machinery provide an abundance of potential sites at which Ca^{++} might exert a regulatory effect on biochemical processes.
3. The soma membranes of many gastropod neurons have been found to have $g_{K(Ca)}$ channels that exert important influences on the integrative properties of a cell (Lux; Meech; both this volume).
4. There is evidence that the voltage-dependent Ca^{++} channels themselves may be regulated by [Ca^{++}]$_i$ (Tillotson, this volume).

To study the processes that determine [Ca^{++}]$_i$, as well as the intracellular effects of Ca^{++}, one would like a method that provides a good degree of both temporal and spatial resolution in the measurement of [Ca^{++}]$_i$. This is particularly important because the intracellular actions of Ca^{++} on the properties of excitable membranes are expressed with a time course of several milliseconds, and spatial gradients of [Ca^{++}]$_i$ may be quite important due to limited diffusion of

The report of this presentation was prepared by J. Strong.

the ion and the nonhomogeneous distribution of various types of Ca^{++} binding sites within the soma.

Here, experiments are described in which voltage-clamp techniques and intracellular Ca^{++} indicators have been used to study the temporal and spatial variations of $[Ca^{++}]_i$ that result when the gastropod soma membrane is depolarized. A simple model capable of explaining many features of the results is presented, and the qualitative behavior and predictions of the model are explored.

A SIMPLE MODEL OF Ca^{++} REGULATION

• A priori, one might expect the neuron's response to a Ca^{++} load to be a quite complex process. The geometry is complicated by the fact that the soma membrane is deeply infolded, giving rise to long, narrow invaginations, and the cytoplasm is full of discrete organelles, microfilaments, etc., which are nonuniformly distributed. The cell may have several systems for sequestering and eliminating Ca^{++} (see Brinley, this volume). However, the data presented below can be explained reasonably well by a model in which many of these potential complications are ignored. The essential features of the model are:

1. The cell is assumed to be a simple sphere.
2. The plasma membrane is assumed to contain pumps that extrude Ca^{++} from the cell at a rate directly proportional to the $[Ca^{++}]_i$ just beneath the membrane.
3. The Ca^{++} buffers of the cell, including any injected Ca^{++} indicators, are represented as a single, uniformly distributed buffer system, and the fraction of buffer that is bound to Ca^{++} is assumed to be directly proportional to $[Ca^{++}]_i$.
4. Ca^{++} enters the cell through voltage-dependent channels in the plasma membrane; characteristics of this current are obtained directly from voltage-clamp data.
5. Using the above assumptions, the $[Ca^{++}]_i$ following a depolarization-induced Ca^{++} influx is computed numerically as a function of time and of distance along a radial axis toward the center of the sphere. For this computation, the interior of the cell is approximated as a series of thin, concentric shells between which Ca^{++} diffuses in a buffer-limited fashion. The diffusion coefficient of free Ca^{++} is assumed to have the values appropriate to dilute aqueous solution, 6.4×10^{-6} cm²/sec (Hodgkin and Keynes 1957). Binding reactions are assumed to be always in equilibrium. Free parameters in the model are the proportionality constants that relate $[Ca^{++}]_i$ to the pump rate and to the amount of Ca^{++} bound by the buffer.

The values assigned to these parameters were chosen to give the best fit of the model to Ca^{++} transients measured using the Ca^{++} indicator arsenazo III. Two features of arsenazo III make it a useful indicator for the studies described here: (1) the dye-Ca^{++} reaction is very rapid, and (2) it yields an absorbance-change signal that is a nearly linear function of Ca^{++} concentration over the range of physiological [Ca^{++}]$_i$.

EXPERIMENTAL JUSTIFICATION FOR A LINEAR MODEL

• It might be expected that the Ca^{++} pump and Ca^{++} buffers would show some sort of saturation phenomena; yet, in the model, it has been assumed that these are acting in a linear fashion. Two different experiments indicate that a linear model is a reasonable first approximation. One such experiment is depicted in Figure 1, which shows the decline in [Ca^{++}]$_i$ measured by the change in arsenazo-III absorbance following 500-msec depolarizations to various V_ms. As the depolarizing pulse is increased from −20 mV to +30 mV, the Ca^{++} influx during the pulse is increased by a factor of 10 or more. Yet the relaxation of [Ca^{++}]$_i$ following the depolarization has approximately the same wave form in each case. This result is explained most simply by assuming that all of the Ca^{++}-removal mechanisms, operating together, behave as a linear system. (The rise of [Ca^{++}]$_i$ during the depolarizing pulse, not shown clearly in Fig. 1, was, in each case, a simple, ramplike rise.)

A second observation that justifies using a linear model for Ca^{++} removal comes from an experiment in which Ca^{++} influx was compared to the maximum buildup of [Ca^{++}]$_i$ for a series of depolarizing steps. Peak arsenazo signals were recorded for 500-msec depolarizing pulses from −20 mV to +20 mV. These signals then were compared to the corresponding values of Ca^{++} influx, which were calculated from Ca^{++} tail currents (I_t) measured by the differential-pair method (see Stevens, this volume) 15 msec after the start of each pulse. A linear relationship was found between the buildup of [Ca^{++}]$_i$ and Ca^{++} influx over this range of V_m. If the Ca^{++} buffers or pumps were operating near saturation, this proportionality would not be predicted; instead, disproportionately larger rises in [Ca^{++}]$_i$ would be seen at the larger depolarizations, where the Ca^{++} influx is greatest.

TWO TYPES OF Ca^{++}-INDICATOR RESPONSES ARE PREDICTED BY THE MODEL

• Figure 2A shows the predicted [Ca^{++}]$_i$ transients in response to a 300-msec voltage-clamp depolarization to +15 mV. Each different

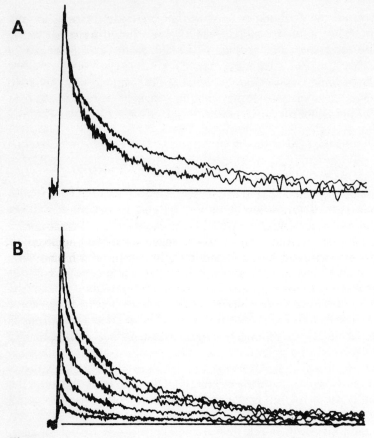

Figure 1
Time course of $[Ca^{++}]_i$ measured as differential absorbance change
of dye arsenazo III at 660 nm and 690 nm, using a chopped-beam
microspectrophotometer. Signals were elicited by 500-msec voltage-clamp pulses to V_ms specified from a V_h of −45 mV. These
data were obtained from a *Helix aspersa* cell loaded with approximately 0.2 mM dye. Full-scale time is 35 sec in both panels. (*A*)
Responses to −10-mV and +30-mV pulses normalized to equal
peak amplitudes. Recovery after the +30-mV pulse is somewhat
slower than that after the −10-mV pulse. (*B*) Responses to pulses
of +30 mV, +20 mV, +10 mV, 0 mV, −10 mV, and −20 mV
plotted with a fixed vertical scale factor. The −20-mV and −10-
mV pulse data are averages of four responses and the 0-mV data
are an average of two responses.

trace represents the time course of changes in $[Ca^{++}]_i$ at a different
distance from the plasma membrane—0.2 μm for the largest response
and 10.8 μm for the smallest. Note that the amplitude of the $[Ca^{++}]_i$
transient decreases rapidly with depth below the surface, although

there is a small but prolonged increase that gradually propagates into the center. In this simulation, cytoplasmic Ca^{++} buffering is represented by assuming that the fraction of Ca bound is 0.99. This value approximates the Ca^{++}-buffering behavior of extruded squid axoplasm described by Baker and Schlaepfer (1975). A linear rate constant governing active efflux of Ca^{++} at the surface membrane was adjusted to optimize the fit of theoretical solutions to experimentally measured arsenazo-III responses. The value finally adopted, 3.2×10^{-3} cm/sec, is within the range characteristic of direct experimental measurements for squid axon Ca^{++} efflux sensitivity to [Ca^{++}]$_i$ (see Blaustein 1976; DiPolo 1976).

Having constrained all parameters by reconstruction of the arsenazo-III response, it is possible to test the model by its ability to predict the response of an indicator with a very different Ca^{++}-sensitivity function, such as the photoprotein aequorin. Figure 2A also shows the computed optical signals one would expect to measure from a cell injected with a linear Ca^{++} indicator (arsenazo III) or with a nonlinear indicator (aequorin). For this prediction, photoemission by aequorin was approximated by a 2.5 power law dependence on [Ca^{++}] (Allen et al. 1977). A similar set of computed predictions is shown in Figure 2B, in which the response to a train of pulses to +15 mV is shown.

Figure 2C shows experimental data obtained under conditions like those simulated in Figure 2B. Note that the model reproduces the form of both indicator responses quite well. In particular, the model correctly predicts that the nonlinear indicator, aequorin, but not arsenazo III, will show a larger incremental response to later pulses. Aequorin measurements such as those shown in Figure 2C have been used as evidence for facilitation of the voltage-dependent g$_{Ca}$; the arsenazo measurements and the model argue against this explanation and attribute the apparent facilitation of Ca^{++} entry seen with aequorin to the nonlinear properties of that indicator (Smith and Zucker 1980).

THE MODEL PREDICTS MEASURED Ca^{++}-DEPENDENT K$^+$ CONDUCTANCE

• Figure 3 shows the slow decay of Ca^{++}-activated K$^+$ I_ts seen in a bursting pacemaker neuron in *Tritonia diomedia* following depolarizing voltage-clamp steps of various durations. The data have been fit using a preliminary, more generalized version of the model described above, in which cytoplasmic Ca^{++} buffering could be represented by a class of saturable, one-to-one reaction sites (see Smith 1978). Such a model is not strictly a linear one, but for the conditions simulated here, deviations from small-signal linear behavior are very

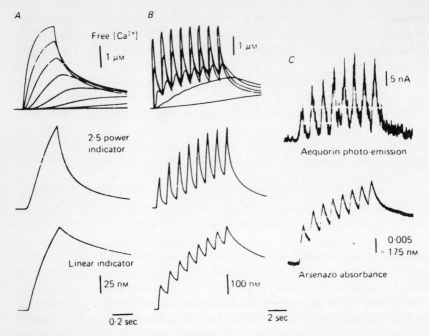

Figure 2
(A, top) Prediction of [Ca^{++}]$_i$ transients during and after a 300-msec voltage-clamped depolarization to +15 mV at depths below the plasma membrane of 0.2 μm, 0.6 μm, 1.2 μm, 1.8 μm, 2.8 μm, 4.3 μm, 6.5 μm, and 10.8 μm. (Middle) Predicted time course of photoemission response of aequorin-loaded soma. (Bottom) Predicted time course of absorbance-change signal in soma loaded with arsenazo III. (B, top) Prediction of [Ca^{++}]$_i$ transient during a train of eight 300-msec pulses to +15 mV, delivered at 1-sec intervals. The traces represent [Ca^{++}]$_i$ at depths of 0.2 μm, 1.2 μm, 2.8 μm, 6.5 μm, and 18.8 μm. (Middle) Predicted aequorin response to train of pulses. (Bottom) Predicted arsenazo-III response to train of pulses. (C) Experimental data obtained from two different *Aplysia* L2 cells under the conditions simulated in B. (Reprinted, with permission, from Smith and Zucker 1980.)

minor. An additional assumption used was that the g$_{K(Ca)}$ is directly proportional to [Ca^{++}]$_i$ level near the membrane. The model, with the addition of this simple assumption, provides a good fit to g$_{K(Ca)}$s observed over a wide range of depolarizing pulse durations.

MEASUREMENT OF SPATIAL GRADIENTS OF Ca^{++} IN A SINGLE CELL

• So far, it has been shown that a relatively simple model of Ca^{++} metabolism can predict, with a few free parameters, the observed

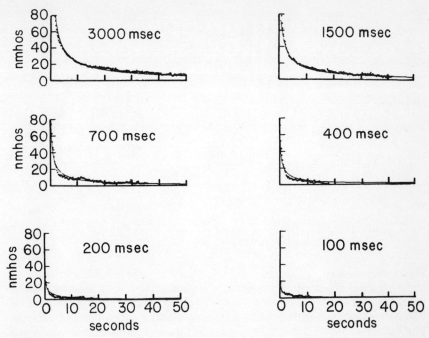

Figure 3
Experimentally measured Ca^{++}-activated K$^+$ I_ts (data points) and theoretical predictions of submembrane [Ca^{++}]$_i$ (solid curves) for *Tritonia* bursting pacemaker neurons at a negative V_h after voltage-clamp pulses to +2 mV for the durations indicated. The theoretical curves were scaled in each frame by the same factor (0.067 nmho/nM), and, thus, both the individual conductance wave forms and the progressive activation with increases in pulse duration are predicted accurately by the model (Smith 1978).

responses of two Ca^{++} indicators to a Ca^{++} influx and the time course of decay of the $g_{K(Ca)}$. In Figure 2 A and B it can be seen that the model predicts very steep spatial gradients of [Ca^{++}]$_i$ following an influx of Ca^{++} through the plasma membrane. Direct measurement of [Ca^{++}]$_i$ gradients, therefore, would provide a further test of the model.

The results of preliminary attempts to achieve spatial resolution of the soma Ca^{++} transients are shown in Figures 4 and 5A. A light-pipe with a diameter considerably smaller than the cell diameter is used to collect the transmitted light. The pipe can be positioned to collect light preferentially from either the center or edge of the cell. The edge signal is presumed to be more heavily weighted with signals originating near the membrane. The arsenazo signals recorded in the two positions as the membrane was depolarized to 0 mV for 5 sec are shown in Figure 4A. Notice that the signal recorded at the edge has a sharper inflection at its peak and a more rapid decline than the center signal. In Figure 4B, the two signals have been scaled

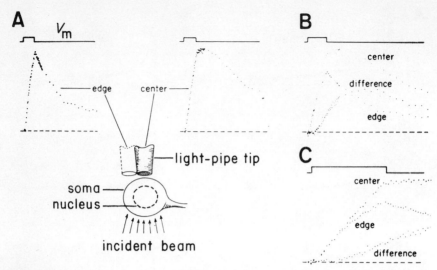

Figure 4
Demonstration of a method for spatially resolving the $[Ca^{++}]_i$ transient within the soma of a *Helix* neuron loaded with approximately 0.6 mM arsenazo III. The data shown are differential absorbance signals at 660 nm and 690 nm. The voltage-clamped potential trajectories indicated in each case correspond to a 5-sec depolarization to 0 mV from a V_h of −45 mV. (A) $[Ca^{++}]_i$ signals recorded during identical depolarizations at the two different light-pipe positions indicated in the diagram. (B) The same data as in A, normalized as described in the text, and the difference between the resulting curves plotted on a common axis. (C) The same data as in B plotted on an expanded time scale to better reveal initial rising phases used to normalize center and edge wave forms.

to have identical initial rates of increase. At the very beginning of the depolarization, both the edge and the center signals should reflect $[Ca^{++}]_i$ changes close to the membrane, since Ca^{++} has not had time to diffuse very far into the cell. Amplitude differences between the two signals at this early time should just reflect differences in path length through peripheral cytoplasm, and the later difference between the two scaled signals thus should be proportional to that portion of the center signal that originates in the interior of the cell. As Figure 4B shows, this difference signal rises quite slowly, continuing to rise for some seconds after the end of the depolarization, and then begins an even more gradual decline. This difference signal may provide an approximation of Ca^{++} signals originating in the nucleus, which occupies most of the soma interior.

Figure 5A depicts the outcome of a similar experiment in which somewhat different results were obtained. The difference between center and edge signals is less profound than in Figure 4, although the center signal still shows a slower recovery time course. A lower

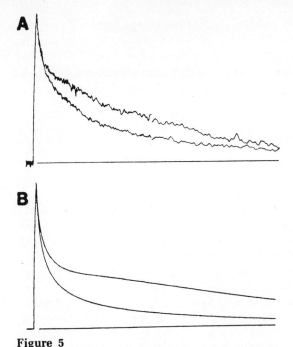

Figure 5
(A) Data from an experiment similar to that
shown in Fig. 4, except that this *Helix* cell was
filled only to about 0.25 mM with arsenazo III.
Center signal (*top trace*); edge signal (*bottom
trace*). (B) Prediction of indicator-response wave
forms appropriate to edge and center light-pipe
positions (see diagram, Fig. 4) for conditions
similar to those imposed experimentally in Fig.
5A. The curve for the center prediction (*top
trace*) corresponds to a spatially averaged [Ca^{++}]$_i$
transient within a cylindrical section through
the center of the simulated soma sphere, and the
curve for the edge prediction (*bottom trace*) corresponds to the average transient for a peripheral layer of the sphere.

dye concentration was employed in the experiment of Figure 5. Since
the dye itself acts to buffer Ca^{++}, the additional buffering in the
experiment of Figure 4 may have acted to retard diffusion and thus
exaggerate spatial gradients. It is possible that Ca^{++} buffering by
arsenazo III may increase in a greater than direct proportion to dye
concentration (see Thomas 1979). Figure 5B shows that a linear Ca^{++}
transient model as discussed above can account for the general form
of differences observed between center and edge signals as in Figure
5A. Reconstruction of arsenazo responses from cells of *Helix* (a terres-

trial species with a hemolymph of relatively low solute concentration), however, has required rather different parameter values than those used for the reconstruction of marine gastropod $[Ca^{++}]_i$ transients. For the preliminary calculations represented in Figure 5B, the fraction of cytoplasmic Ca that is bound was set at only 0.95.

SUMMARY AND FUTURE QUESTIONS

• The simple model of Ca^{++} metabolism presented above, which is based on the assumption of a Ca^{++} pump in the plasma membrane and buffer-limited diffusion of Ca^{++} within the cytosol, has successfully predicted four types of experimental observations: (1) the aequorin response to a voltage-induced Ca^{++} influx, (2) the arsenazo-III response to a voltage-induced Ca^{++} influx, (3) the kinetics of decay of the $g_{K(Ca)}$, and (4) the spatial distribution of Ca^{++} within the cell.

Improvements in the methodology for measuring $[Ca^{++}]_i$ gradients within a single cell would be of general interest, completely independent of any model. One such improvement that is planned is to measure the transmitted light intensity in a real-image plane, rather than measuring it with a photodetector placed directly over the cell. This method offers several advantages: The small light-pipe may be placed with greater accuracy over the magnified image, there will be room to place several light-pipes over different parts of the cell simultaneously, light-scattering problems caused by the close proximity of the light-pipe to the cell will be eliminated, and the improved optics will allow localization of discrete elements of the cytoplasm such as the nucleus. Such an improved methodology would provide a powerful tool for studying a variety of interesting processes. One could study the buffering, diffusion, regulation, and extrusion of intracellular Ca^{++} in an essentially intact cell under physiological conditions. By making V_m and ionic-current recordings at the same time that one measures $[Ca^{++}]_i$ levels, it may also be possible to elucidate the relationships between Ca^{++} regulatory processes and the excitable properties of the nerve cell membrane.

REFERENCES

Allen, D.G., J.R. Blinks, and F.G. Prendergast. 1977. Aequorin luminescence: Relation of light emission to calcium concentration—A calcium-independent component. *Science* **195**: 996.

Baker, P.F. and W. Schlaepfer. 1975. Calcium uptake by axoplasm extruded from giant axons of *Loligo*. *J. Physiol. (Lond.)* **239**: 37P.

Blaustein, M.P. 1976. The ins and outs of calcium transport in squid axons: Internal and external ion activation of calcium efflux. *Fed. Proc.* **35**: 2574.

DiPolo, R. 1976. The influence of nucleotides on calcium fluxes. *Fed. Proc.* **35:** 2579.

Hodgkin, A.L. and R.D. Keynes. 1957. Movements of labelled calcium in squid giant axons. *J. Physiol. (Lond.)* **138:** 253.

Smith, S.J. 1978. "The mechanism of bursting pacemaker activity in neurons of the mollusc *Tritonia diomedia.*" Ph.D. thesis, University of Washington, Seattle.

Smith, S.J. and R.S. Zucker. 1980. Aequorin response facilitation and intracellular calcium accumulation in molluscan neurones. *J. Physiol. (Lond.)* **300:** 167.

Thomas, M.V. 1979. Arsenazo III forms 2:1 complexes with Ca and 1:1 complexes with Mg under physiological conditions. *Biophys. J.* **25:** 541.

Ca^{++}-activated K$^+$ Conductance

Based on a presentation by

ROBERT W. MEECH

Department of Physiology
University of Utah
Salt Lake City, Utah 84112

• The observation that revealed the existence of $g_{K(Ca)}$ in nerve membranes was a simple one: Microinjection of Ca^{++} into the nerve cell body of an *Aplysia* neuron was found to hyperpolarize the cell membrane (Meech and Strumwasser 1970; Meech 1972). Three types of evidence indicate that there is a specific increase in g_K:

1. The hyperpolarization is associated with a fall in membrane resistance.
2. The reversal potential of the response depends on the concentration of K$^+$ in the bathing medium in a manner predicted by the Nernst equation for K$^+$.
3. The hyperpolarization is abolished by TEA$^+$, which is known to block K$^+$ channels in many different preparations.

It was shown subsequently that $g_{K(Ca)}$ channels can also be activated by depolarizing pulses or action potentials that open Ca^{++} channels and cause an influx of Ca^{++} through the membrane into the cytoplasm (Meech 1974a,b; Meech and Standen 1975; Heyer and Lux 1976; Thompson 1977). In many molluscan neurons, there is a prolonged increase in g_K after a train of spikes (Brodwick and Junge 1972; Moreton 1972) which is associated with a build-up of Ca^{++} inside the cell (Stinnakre and Tauc 1973; Thomas and Gorman 1977). In *Aplysia*, the conductance increase can be abolished by injecting the Ca^{++}-chelating agent EGTA (Meech 1974b). Thus, it seems likely that the K$^+$ conductance increase is a Ca^{++}-mediated effect, if one assumes that the EGTA buffers the transmembrane Ca^{++} influx before it can activate K$^+$ channels.

The report of this presentation was prepared by G. Yellen.

A similar long-lasting g_K increase is observed in *Helix* neurons following repetitive depolarizing pulses under voltage-clamp control (Meech 1974a). During a single pulse, $I_{K(Ca)}$ can be partially separated from other components by a difference technique (Meech and Standen 1975). The slowly developing outward current in response to the same depolarizing step is measured first in normal Ca^{++}-containing saline and then in saline in which the $CaCl_2$ has been replaced by $MgCl_2$. Subtracting the current recorded in Mg^{++} from that recorded in Ca^{++} then provides a measure of $I_{K(Ca)}$. The effect of the nominally Ca^{++}-free solution is to reduce significantly the size of the outward current over a wide range of command potentials. Figure 1A shows this procedure for outward current measured at a specific time (80 msec) after the beginning of the depolarizing pulse. The Ca^{++}-dependent component of I_K, separated by subtraction, is shown in Figure 1B. The small size of $I_{K(Ca)}$ during large depolarizing steps probably is due to the reduced entry of Ca^{++} at these V_ms, as demonstrated by the aequorin experiments of Baker et al. (1971) and Lux and Heyer (1977).

The g_Ks activated by Ca^{++} injection and by Ca^{++} influx across the cell membrane are both referred to as $g_{K(Ca)}$. The evidence that g_K activated by a depolarization-induced Ca^{++} influx is the same conductance as that activated by Ca^{++} microinjection can be summarized briefly.

1. The g_K activated following Ca^{++} influx across the cell membrane is abolished by the microinjection of EGTA into the cell cytoplasm (Meech 1978; Connor 1979).
2. There is little $I_{K(Ca)}$ at depolarizations close to E_{Ca} (about −165 mV [Stinnakre and Tauc 1973]), where the net increase in intracellular Ca^{++} would be expected to be small (Meech and Standen 1975).
3. The I_K that develops slowly in response to depolarization of a perfused neuron bathed in nominally Ca^{++}-free saline containing 40 mM TEA is potentiated when the $[Ca^{++}]_i$ is raised from 1×10^{-7} M to 6×10^{-7} M (Doroshenko et al. 1979).

On the other hand, this idea is not accepted by Lux and his coworkers (see Lux, this volume), who have argued that the two conductance mechanisms are not equivalent. They have demonstrated that relatively large loads of injected Ca^{++} cause a depression of the slowly developing outward current. This effect might be explained, however, by very high levels of $[Ca^{++}]_i$ having a nonspecific, unphysiological effect on the cell interior in addition to its normal actions on $g_{K(Ca)}$ channels (Tobias and Bryant 1955). To interpret more complex experiments, one must understand in detail how intracellular Ca^{++} is buffered under specific experimental conditions. This presentation describes how microinjected CaCl$_2$ is buffered by the cell. In addition, there is an account of the functional properties of the Ca^{++} receptors within the cell that mediate the effect of Ca^{++} injection on g$_K$. How tightly does Ca^{++} bind to the receptor, how selective is the binding site for Ca^{++}, and how many Ca^{++} must bind to open a single K$^+$ channel? The experiments described (see Meech and Thomas 1980) were performed, in most cases, on cell A in the right parietal ganglion of *Helix aspersa*. Standard microinjection and ion-specific microelectrode techniques were employed (see Thomas 1978; Meech 1981).

MICROINJECTED Ca^{++} IS BUFFERED MAINLY BY MITOCHONDRIA

• As in squid axons (see Brinley, this volume), Ca^{++} is bound by both mitochondrial and nonmitochondrial fractions of the cytoplasm. The relative importance of the sites depends upon the experimental conditions. The main findings in *Helix* neurons are:

1. In the absence of mitochondrial binding, the large loads of Ca^{++} achieved by microinjection are buffered relatively slowly.

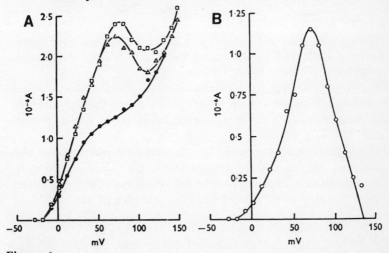

Figure 1
Separation of $I_{K(Ca)}$ by difference technique. (A) Effect of Ca^{++}-free saline on the relationship between the outward I_m (ordinate) and V_m (abscissa). I_m measured 80 msec after the beginning of command pulse. Normal (□); after 3 min in Ca^{++}-free saline (•); 4–5 min after return to normal saline (Δ). (B) Ca^{++}-dependent component of current shown in A, obtained by subtraction of currents recorded in Ca^{++}-free saline from those in normal saline. V_h, −48 mV. (Reprinted, with permission, from Meech and Standen 1975.)

2. After a Ca^{++} load, the metabolic inhibitor CCmP produces a hyperpolarization that is not seen in uninjected cells. In isolated mitochondria, such agents are known to cause the release of previously sequestered Ca^{++} (Vasington and Murphy 1962).

3. CCmP does not hyperpolarize ruthenium-red-injected cells, even when they have been preloaded with as much as 24 mEq Ca^{++}/liter before CCmP application. The dye ruthenium red inhibits Ca^{++} uptake by isolated mitochondria without preventing its release (Vasington et al. 1972).

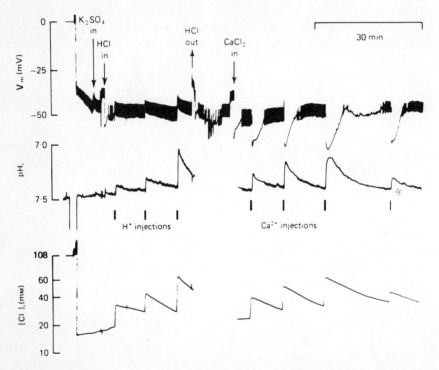

Figure 2
Pen-recording of an experiment to compare the effects of HCl and $CaCl_2$ injection on pH_i. Five microelectrodes were inserted into a single cell in the following order: (1) Cl^--sensitive electrode, (2) pH-sensitive glass electrode, (3) V_m-recording micropipette, (4) current-injecting, K_2SO_4-filled micropipette, (5) pressure-injection micropipette. For Ca^{++} injection, the injection pipette was filled with a solution containing 0.1 M $CaCl_2$ and 0.1 M KCl; for acid injection it was filled with 0.1 M HCl and 0.2 M KCl. Insertion of the HCl injection pipette (↓) was followed by three separate injections (I). The HCl injection pipette was then replaced by one filled with $CaCl_2$, and four injections of Ca^{++} followed. The gap in the pH_i and $[Cl^-]_i$ traces corresponds to the period during which the injection pipettes were exchanged. The cell was bathed throughout in saline equilibrated with 2.2% CO_2. (Reprinted, with permission, from Meech and Thomas 1980.)

4. Microinjection of CaCl₂ is accompanied by a fall in pH_i. The Ca^{++}-H^+ exchange is approximately one for one (see Figs. 2 and 3), the same as observed in isolated mitochondria (Chappell et al. 1962).

Taken together, this evidence strongly suggests that a major part of the Ca^{++} injected under the experimental conditions described is taken up into mitochondria (Meech and Thomas 1980).

THE CA++ RECEPTOR IS TROPONINLIKE

• The Ca^{++} receptor for $g_{K(Ca)}$ is in several respects like troponin C, the small Ca^{++}-binding protein involved in coupling intracellular Ca^{++} release to fiber contraction in muscle cells. Troponin C has a structure similar to that of the ubiquitous Ca^{++}-binding protein known variously as calmodulin or Ca^{++}-dependent regulator (Wang et al. 1974; Vanaman et al. 1977). In the absence of Mg^{++}, both troponin C and calmodulin bind Ca^{++} at four sites (Potter and Gergely 1975; Wolff et al. 1977).

There are three lines of evidence that indicate a similarity between these Ca^{++}-binding proteins and the Ca^{++} receptor for $g_{K(Ca)}$.

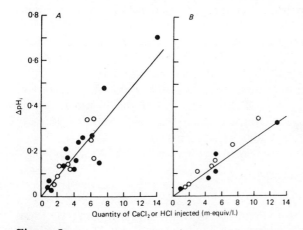

Figure 3

The change in pH_i (ΔpH_i) produced by injecting known quantities of HCl or CaCl₂. Cells equilibrated with 1% (A) and 2.2% (B) CO₂. Quantity of HCl (•) or CaCl₂ (o) injected (abscissa); change in pH_i on the same scale for each group (ordinate). The quantity of HCl or CaCl₂ injected was calculated from the change in $[Cl^-]_i$ shown by the Cl^--sensitive electrode. The line in each case represents the average of the measurements of buffering capacity made on each group of cells. (Reprinted, with permission, from Meech and Thomas 1980.)

Both the Ca^{++} Receptor and Troponin C Bind Micromolar Levels of Ca^{++}

In the original microinjection experiments, it was necessary to inject millimolar quantities of $CaCl_2$ to produce a hyperpolarizing response (Meech 1972, 1974a). If Ca-EGTA buffers are used instead of $CaCl_2$, micromolar levels of Ca^{++} can produce a response (Meech 1974a). The injected Ca-EGTA does not, in all probability, totally buffer the intracellular Ca^{++} level, but it is safe to say that the $[Ca^{++}]_i$ does not rise above the buffered level in the injected solution. This means that the Ca^{++} receptor must be sensitive to $[Ca^{++}]_i$ at least as low as the micromolar range. For troponin C, the dissociation constant (K_D) of the active Ca^{++}-binding sites in the presence of 2 mM Mg^{++} is 5×10^{-6} M (Potter and Gergely 1975).

The Ion Selectivity of the Ca^{++} Receptor Is Like That of Troponin C

The divalent cations that compete most effectively with Ca^{++} for binding to troponin C (Fuchs 1971) are the same as those that produce a rapid increase in g_K when microinjected into snail neurons; they are Pb^{++}, Sr^{++}, Hg^{++}, Ca^{++}, Cd^{++}, and, possibly, Mn^{++} (see Meech 1976). In both cases, selectivity seems to be based on size, with those ions having an ionic radius of about 1 Å being bound most effectively. This result has been confirmed by Gorman and Hermann (1979), who used an iontophoretic method of injection. Unfortunately, it is not yet possible to quantitate fully the differences in effectiveness because there is insufficient information about the capacity of the cell cytoplasm to sequester the different divalent ions tested. Furthermore, microinjection of $BaCl_2$ can cause the intracellular release of Ca^{++} (Meech and Thomas 1980). Whether this is a property common to other divalent ions is not known.

The Ca^{++} Receptor Binds At Least 3 Ca^{++}

The evidence that at least 3 Ca^{++} must bind simultaneously to activate $I_{K(Ca)}$ comes from the shape of the dose-response curve of membrane hyperpolarization vs injected Ca^{++} (Meech and Thomas 1980).

In determining this dose-response relation, the $[Ca^{++}]$ at the inner surface of the cell membrane is assumed to be related linearly to the amount of Ca^{++} injected. This is based on two findings: (1) In ruthenium-red-injected cells, Ca^{++} sequestration apparently is operating at a constant velocity, and (2) the relationship between the time to peak hyperpolarization and the amount of Ca^{++} injected is linear. Taken together, these results indicate that the increase in free Ca^{++} at the time of maximum hyperpolarization must be a fixed proportion of the quantity injected (probably about 0.1%, [Meech 1974a]). Figure 4A

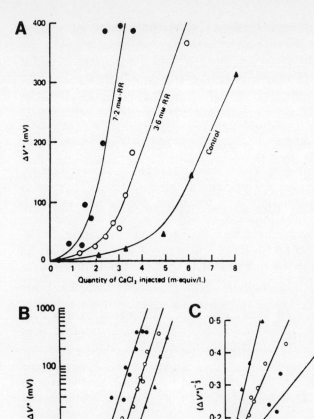

Figure 4

Graphs showing the effect of different concentrations of inject-ed Ca^{++} on the size of the corrected hyperpolarization, ΔV^*, for a control cell (▲) and cells injected to a final concentration of 3.6 mM (o) and 7.2 mM (•) ruthenium red (RR). (A) Quantity of $CaCl_2$ injected (abscissa); ΔV^*, the Ca^{++}-induced voltage change (ΔV_m) multiplied by $[(V_m - E_{rev})/(V_m - E_{rev} - \Delta V_m)]$ (ordi-nate). ΔV^* is proportional to the $g_{K(Ca)}$ (Martin 1955; Glitsch and Pott 1978). (B) Same data plotted on logarithmic coordi-nates. Lines of slope 3 aligned by eye to fit the points. (C) Lineweaver-Burke plot of same data. Reciprocal of the concen-tration of the Ca^{++} injected (abscissa); reciprocal of the cube root of ΔV^* (ordinate). The point $1/[CaCl_2] = 3.03$; $[(\Delta V_m)]^{-1/3} = 0.93$ has been plotted as 1.515, 0.465. (Reprinted, with permis-sion, from Meech and Thomas 1980.)

shows the hyperpolarizing responses to different amounts of injected Ca^{++} in cells with and without ruthenium red injection. The responses have been corrected for the fact that the driving force for hyperpolarization decreases as V_m approaches the E_{rev} for $g_{K(Ca)}$. These corrected hyperpolarizing responses, which are proportional to $g_{K(Ca)}$, increase approximately with the third power of the Ca^{++}load. This result suggests that the combined action of at least 3 Ca^{++} is necessary to activate $g_{K(Ca)}$ (the fit to the third-power relation is shown in different ways in Fig. 4, B and C).

Gorman and Hermann (1979) have reported a linear dose-response curve for $I_{K(Ca)}$ as a function of injected Ca^{++}. They used iontophoretically injected Ca^{++} and measured the current response under voltage clamp. The largest Ca^{++} injection used in their measurements was approximately 0.92 mEq/liter (see Gorman and Hermann 1979). This is well within the apparently linear portion of the dose-response curve shown in Figure 4A, so there would seem to be no essential disagreement with the results of Meech and Thomas (1980).

The P_K of the red cell membrane is also increased by a rise in $[Ca^{++}]_i$. In this system, the number of equivalent Ca^{++}-binding sites that must be filled before K^+ transport can occur is a minimum of two (Simons 1976).

For troponin C, Potter and Gergely (1975) suggest that binding of Ca^{++} to two sites is sufficient to produce full activation of the myofibrillar ATPase, although involvement of the other sites cannot be excluded. On the other hand, the Ca^{++}-dependent regulator from brain binds Ca^{++} in the presence of 1 mM Mg^{++} at a single class of sites in a molar ratio of 3:1, with an apparent K_D of 3×10^{-6} M (Wolff et al. 1977; Wolff and Brostrom 1979). It is not clear whether or not binding of all 3 Ca^{++} is required to produce the active conformational state of the protein.

DISCUSSION

● Future work on the mechanism of the $g_{K(Ca)}$ channel will need to address the specific question of how closely the structure of the receptor for activating this channel resembles other Ca^{++}-binding proteins. Many other mechanistic questions remain. Is the receptor free in the cytoplasm or is it membrane-bound? How does it activate the K^+ channel, and what are the ion selectivity properties of this channel? An important problem is to distinguish the direct effects of internal Ca^{++} from indirect effects mediated by changes in pH_i, because K^+ channels appear to be sensitive to small changes in pH_i (Meech 1979; Wanke et al. 1979). A way around this problem would be to use perfused neurons (e.g., see Doroshenko et al. 1979), but it is possible that the Ca^{++}-binding receptor is not membrane-bound—at least in its free state (see Watterson et al. 1976; Kakiuchi et al. 1978). If so, prolonged intracellular perfusion may progressively remove the receptor from the cytoplasm. In spite of the difficulties, there may be advantages to using intact cells, and it is encouraging to find that microinjection causes little disturbance of the cell cytoplasm (Nicaise and Meech 1980).

REFERENCES

Baker, P.F., A.L. Hodgkin, and E.B. Ridgway. 1971. Depolarization and calcium entry in squid giant axons. *J. Physiol. (Lond.)* **218:** 709.

Brodwick, M.S. and D. Junge. 1972. Post-stimulus hyperpolarization and slow potassium conductance increase in *Aplysia* giant neurone. *J. Physiol. (Lond.)* **223:** 549.

Chappell, J.B., G.D. Greville, and K.E. Bicknell. 1962. Stimulation of respiration of isolated mitochondria by manganese ions. *Biochem. J.* **84:** 61P.

Connor, J.A. 1979. Calcium current in molluscan neurones: Measurement under conditions which maximize its visibility. *J. Physiol. (Lond.)* **24:** 41.

Doroshenko, P.A., P.G. Kostyuk, and A.Y. Tsyndrenko. 1979. Investigation of the TEA-resistant outward current in the somatic membrane of perfused nerve cells. *Neurophysiology* **11:** 341.

Fuchs, F. 1971. Ion exchange properties of the calcium receptor site of troponin. *Biochim. Biophys. Acta* **245:** 221.

Glitsch, H.G. and L. Pott. 1978. Effects of acetylcholine and parasympathetic nerve stimulation on membrane potential in quiescent guinea-pig atria. *J. Physiol. (Lond.)* **279:** 655.

Gorman, A.L.F. and A. Hermann. 1979. Internal effects of divalent cations on potassium permeability in molluscan neurones. *J. Physiol. (Lond.)* **296:** 393.

Heyer, C.B. and H.D. Lux. 1976. Control of the delayed outward potassium currents in bursting pace-maker neurones of the snail, *Helix pomatia*. *J. Physiol. (Lond.)* **262:** 349.

Kakiuchi, S., R. Yamazaki, Y. Teshima, K. Uenishi, S. Yasuda, A. Kashiba, K. Sobue, M. Ohshima, and T. Nakajima. 1978. Membrane-bound protein modulator and phosphodiesterase. *Adv. Cyclic Nucleotide Res.* **9:** 253.

Lux, H.D. and C.B. Heyer. 1977. An aequorin study of a facilitating calcium current in bursting pacemaker neurons of *Helix*. *Neuroscience* **2:** 585.

Martin, A.R. 1955. A further study of the statistical composition of the end-plate potential. *J. Physiol. (Lond.)* **130:** 114.

Meech, R.W. 1972. Intracellular calcium injection causes increased potassium conductance in *Aplysia* nerve cells. *Comp. Biochem. Physiol.* **42A:** 493.

————. 1974a. The sensitivity of *Helix aspersa* neurones to injected calcium ions. *J. Physiol. (Lond.)* **237:** 259.

————. 1974b. Calcium influx induces a post-tetanic hyperpolarization in *Aplysia* neurones. *Comp. Biochem. Physiol.* **48A:** 387.

————. 1976. Intracellular calcium and the control of membrane permeability. *Symp. Soc. Exp. Biol.* **30:** 161.

————. 1978. Calcium-dependent potassium activation in nervous tissues. *Annu. Rev. Biophys. Bioeng.* **7:** 1.

————. 1979. Membrane potential oscillations in molluscan "burster" neurones. *J. Exp. Biol.* **81:** 93.

————. 1981. Microinjection. In *Techniques in cellular physiology* (ed. P.F. Baker). Elsevier/North-Holland, Amsterdam. (In press.)

Meech, R.W. and N.B. Standen. 1975. Potassium activation in *Helix aspersa* neurones under voltage clamp: A component mediated by calcium influx. *J. Physiol. (Lond.)* **249:** 211.

Meech, R.W. and F. Strumwasser. 1970. Intracellular calcium injection activates potassium conductance in *Aplysia* nerve cells. *Fed. Proc.* **29:** 835.

Meech, R.W. and R.C. Thomas. 1980. Effect of measured calcium chloride injections on the membrane potential and internal pH of snail neurones. *J. Physiol. (Lond.)* **298:** 111.

Moreton, R.B. 1972. Electrophysiology and ionic movements in the central nervous system of the snail, *Helix aspersa. J. Exp. Biol.* **57:** 513.

Nicaise, G. and R.W. Meech. 1980. The effect of pressure injection on the ultrastructure of molluscan neurones. *Brain Res.* **193:** 549.

Potter, J.D. and J. Gergely. 1975. The calcium and magnesium binding sites on troponin and their role in the regulation of myofibrillar adenosine triphosphatase. *J. Biol. Chem.* **250:** 4628.

Simons, T.J.B. 1976. Calcium-dependent potassium exchange in human red cell ghosts. *J. Physiol. (Lond.)* **256:** 227.

Stinnakre, J. and L. Tauc. 1973. Calcium influx in active *Aplysia* neurones detected by injected aequorin. *Nature New Biol.* **242:** 113.

Thomas, M.V. and A.L.F. Gorman. 1977. Internal calcium changes in a bursting pacemaker neuron measured with arsenazo III. *Science* **196:** 531.

Thomas, R.C. 1978. *Ion-sensitive intracellular microelectrodes.* Academic Press, London.

Thompson, S.H. 1977. Three pharmacologically distinct potassium channels in molluscan neurones. *J. Physiol. (Lond.)* **265:** 465.

Tobias, J.M. and S.H. Bryant. 1955. An isolated giant axon preparation from the lobster nerve cord. *J. Cell. Comp. Physiol.* **46:** 163.

Vanaman, T.C., F. Sharief, and D.M. Watterson. 1977. Structural homology between brain modulator protein and muscle TnCs. In *Calcium-binding proteins and calcium functions* (ed. R.H. Wasserman et al.), p. 107. Elsevier/North-Holland, New York.

Vasington, F.D. and J.V. Murphy. 1962. Ca^{++} uptake by rat kidney mitochondria and its dependence on respiration and phosphorylation. *J. Biol. Chem.* **237:** 2670.

Vasington, F.D., P. Gazzoti, R. Tiozzo, and E. Carafoli. 1972. The effect of ruthenium red on Ca^{2+} transport and respiration in rat liver mitochondria. *Biochim. Biophys. Acta* **256:** 43.

Wang, J.H., T.S. Teo, H.C. Ho, and F.C. Stevens. 1975. Bovine heart protein activator of cyclic nucleotide phosphodiesterase. *Adv. Cyclic Nucleotide Res.* **5:** 179.

Wanke, E., E. Carbone, and P.L. Testa. 1979. K$^+$ conductance modified by a titratable group accessible to protons from the intracellular side of the squid axon membrane. *Biophys. J.* **26:** 319.

Watterson, D.M., W.G. Harrelson, Jr., P.M. Keller, F. Sharief, and T.C. Vanaman. 1976. Structural similarities between the Ca^{2+}-dependent regulatory proteins of 3':5'-cyclic nucleotide phosphodiesterase and actomyosin ATPase. *J. Biol. Chem.* **251:** 4501.

Wolff, D.J. and C.O. Brostrom. 1979. Properties and functions of the calcium-dependent regulator protein. *Adv. Cyclic Nucleotide Res.* **11:** 27.

Wolff, D.J., M.A. Brostrom, and C.O. Brostrom. 1977. Divalent cation binding sites of CDR and their role in the regulation of brain cyclic nucleotide metabolism. In *Calcium-binding proteins and calcium function* (ed. R.H. Wasserman et al.), p. 97. Elsevier/North-Holland, New York.

Voltage Dependence of Ca^{++}-activated K$^+$ Conductance

Based on a presentation by

H. DIETER LUX*

Max-Planck-Institut für Psychiatrie
Munich, Federal Republic of Germany 401240

• In molluscan neurons, Meech and Standen (1975) and Heyer and Lux (1976b) have previously identified an outward I_K that depends on the presence of extracellular Ca^{++} and is produced by a depolarization-activated inward I_{Ca}. At present, it is unknown how this K$^+$ current, designated $I_{K(Ca)}$, is linked functionally with membrane I_{Ca}. One proposal has been that $I_{K(Ca)}$ is activated simply by the elevation of [Ca^{++}]$_i$ (Meech and Standen 1975). This hypothesis is supported by correlations between outward current and [Ca^{++}]$_i$, as monitored with Ca^{++} indicators (Eckert and Tillotson 1978; S. Smith, this volume), as well as by the long-term activation of outward current by intracellular injection of Ca^{++} (Meech, this volume).

An alternative and more complex view is that $I_{K(Ca)}$ is dependent on both [Ca^{++}]$_i$ and voltage for activation. This is suggested by the observation that when cell depolarization is terminated by strong hyperpolarization, $I_{K(Ca)}$ (along with other membrane currents) disappears rather quickly (Heyer and Lux 1976a). The rapid decline in $I_{K(Ca)}$ during hyperpolarization does not appear to be compatible with the relatively slow rate with which intracellularly accumulated Ca^{++} equilibrates, as observed with intracellular Ca^{++} indicators (Eckert et al. 1977; Lux and Heyer 1977; Gorman and Thomas 1978; Ahmed and Connor 1979; but cf. S. Smith, this volume). Two kinds of experiments, which are described here, were performed to study the properties of $I_{K(Ca)}$: (1) intracellular injection of Ca^{++} using a rapid-pulse

---

*The work presented here was done in collaboration with G. Hofmeier.
The report of this presentation was prepared by D. Harris and S. Mackey.

105

technique and (2) production of outward I_K by membrane depolarization that activates Ca^{++} channels.

INTRACELLULAR Ca^{++} INJECTIONS

• Experiments were performed on the fast burster cell in the right parietal ganglion of the snail *Helix pomatia* (Heyer and Lux 1976a). Because rapid effects that might be obscured by slow iontophoretic application of Ca^{++} were of primary interest, it was preferable to inject controlled quantities of $CaCl_2$ solution by pressure pulses of tenths of a second according to a method modified from Llinás et al. (1972). $[Ca^{++}]_i$ was monitored by Ca^{++}-sensitive microelectrodes that were fabricated using the basic techniques of Lux and Neher (1973; see also Heinemann et al. 1977). The voltage-clamp arrangement used methods described previously (Heyer and Lux 1976a).

Figure 1 (top) shows the characteristic trace of Ca^{++}-activated membrane currents that occur in response to an injection pulse with

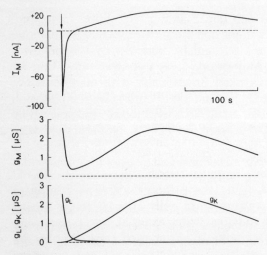

Figure 1
I_m and increased conductance after termination of an injection of $CaCl_2$ into a neuron clamped at -50 mV. Conductance was monitored by hyperpolarizing pulses to -70 mV at 0.2 Hz, and the control value of 0.17 μS is subtracted in the g_m plot. Bottom trace shows time courses of the components g_L and g_K. Calculated equilibrium potentials were -16 mV for g_L and -60 mV for g_K. Both values were determined during their respective peaks.

the cell clamped to a V$_h$ of -50 mV. The response was always bi-phasic—rapid development of an inward current followed by a more prolonged outward current. It was possible to show (Fig. 1) that the total conductance underlying this biphasic current could be fraction-ated into two components—a rapidly rising leakage conductance, g$_L$, and a more slowly rising K$^+$-specific conductance, g$_K$ (Hofmeier and Lux 1980). It is this latter conductance that hitherto has been known as the primary effect of internally injected Ca^{++} (Meech 1972, 1974).

A striking feature of g$_K$ apparent with the pulse-injection tech-nique used in this study is that g$_K$ reaches its maximum long after [Ca^{++}]$_i$ has begun to decline (Hofmeier and Lux 1980). This indicates that Ca^{++} probably is not the direct mediator of g$_K$ opening but is essential for some intermediate step in channel activation. Consistent with this notion are observations suggesting that open K$^+$ channels are in fact partially blocked by additional Ca^{++} (Heyer and Lux 1976b; Hofmeier and Lux 1980).

A crucial property of the g$_{K(Ca)}$ induced by the Ca^{++}-pulse injec-tion is that the underlying P$_K$ is voltage-independent. The peak I-V curve for this g$_{K(Ca)}$ can be fit by a Goldman-Hodgkin-Katz constant permeability curve.

$I_{K(Ca)}$ INDUCED BY MEMBRANE DEPOLARIZATION

• Unlike the increase in P$_K$ resulting from pressure injection of Ca^{++}, the Ca^{++}-activated increase in P$_K$ caused by membrane depolarization shows clear voltage dependence. This strongly suggests that $I_{K(Ca)}$ elicited by depolarization may not be identical with the $I_{K(Ca)}$ ob-served after intracellular injection of Ca^{++}.

Lux and coworkers have isolated specific pacemaker cells (U cells) in *Helix* in which $I_{K(Ca)}$ is the principal carrier of outward current. The early inward current in U cells appears to be carried primarily by Ca^{++}, since it remains when external Na$^+$ is removed. The outward current in these cells is suppressed reversibly by the substitution of either Mg^{++} or Ni^{++} for external Ca^{++} and is depen-dent upon the presence of external Ca^{++} in the submillimolar range. Recordings with extracellular K$^+$-sensitive microelectrodes show that enough K$^+$ is liberated to account quantitatively for the electric charge transferred by Ca^{++}-dependent outward current. The purity of $I_{K(Ca)}$ in these cells makes it possible to analyze its activation time course and voltage dependence. A model based on the voltage depen-dence of the activation can be used to describe quantitatively the basic features of $I_{K(Ca)}$.

$I_{K(Ca)}$ was observed after depolarizations from a V$_h$ of -50 mV (Fig. 2A). The shortest rise times of these currents (10–15 msec) were at V_ms between -10 mV and -20 mV. With increasing depolarizing steps, the time to half-maximum for $I_{K(Ca)}$ increased e-fold for a rise in V_m of about 30 mV. The delay between depolarization and onset of activation of $g_{K(Ca)}$ also showed a pronounced increase with increasing depolarizations. This behavior is quite in contrast to the activation time courses of inward $I_{(Ca)}$ (Kostyuk and Krishtal 1977; Akaike et al. 1978) and intracellular Ca^{++} accumulation. At V_m levels between $+110$ mV and $+130$ mV, slopes of the currents almost paralleled the reference (holding) current for about 100 msec, and time courses could not be determined accurately. No significant decay of currents during voltage steps of up to 400-msec duration was observed.

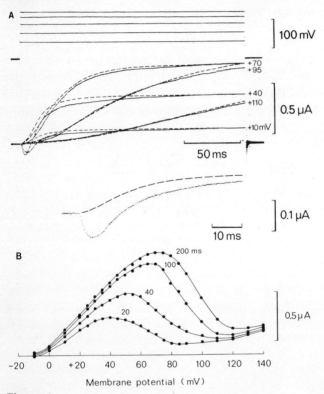

Figure 2
(A) Outward currents activated by steps to different V_m levels (———) and values of $I_{K(Ca)}$ calculated from Eqs. 1–4 (- - - -). (Below) Expanded current trace at $+5$ mV to show early inward current component. (B) Isochronal I-V relationships of outward currents. Measurements at four indicated times reveal the characteristic shift of maxima and minima of total outward currents toward more positive V_ms with increasing time of measurement.

$I_{K(Ca)}$ is characterized by bell-shaped isochronal I-V curves (Fig. 2B). The peaks of these curves move from +30 mV to about +70 mV with an increase in the time of measurement from 30 msec to 200 msec. The bell shape of these I-V curves reflects the progressive slowing of the activation time for $I_{K(Ca)}$ with increased depolarization (Fig. 2A).

A simple mathematical description which is able to account for the voltage dependence of the time course of $I_{K(Ca)}$ is as follows:

$$I_{K(Ca)} = \bar{g}\ A_K^n\ (V_m - E_K) \tag{1}$$

This equation was chosen by analogy to the equation developed for Na$^+$ and K$^+$ channels of squid axons (Hodgkin and Huxley 1952). It

A

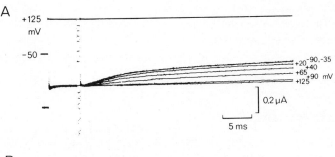

B

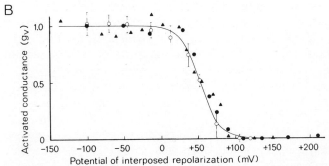

Figure 3

(A) Effect on outward current activation of varying the potential of IR of 0.5-msec duration. Activated outward currents at +125 mV are denoted by the potential level of the individual IRs by which they were produced. (B) Dependence of activated conductance on V_ms of the IRs. Normalized plot of activated conductances measured 50 msec after the IRs. V_ms of the test depolarizations in three U cells were between +110 mV and +125 mV. Line is calculated fitting equation

$$g_v/g_{max} = \{1 + \exp(-\alpha + \beta V_m)/12.5\ \text{mV})\}^{-1}$$

with V_m being the membrane potential of the IR. Curve parameters are $\alpha = 52.5$ mV, $\beta = 1.0$.

includes a maximum conductance $\bar{g}$ and a voltage-dependent activation variable A_K, which changes with first-order kinetics:

$$A_K = 1 - \exp(-t/\tau_V) \tag{2}$$

$V_m - E_K$ denotes the driving force on K^+, with the K^+ equilibrium potential, E_K, being -70 mV. The choice of $n = 2$ resulted in the best fit of currents at all depolarizations. The exponential prolongation of activation time courses with increasing depolarizing steps and the minimal rise time observed for very small depolarizations are both well fit by the form

$$\tau_v = \tau_o[1 + \exp \ (-\alpha + \beta V_m)] \tag{3}$$

with the exponent being a linear function of voltage. The lower limit of the time constant, τ_0, as well as the factor representing the voltage dependence of activation were determined independently from the results of two kinds of experiments: (1) prepulse injection to obtain τ_0 and (2) interposed repolarization (IR) to obtain voltage dependence of activation.

Prepulse Experiments and τ_o

In prepulse experiments, a small depolarizing voltage step that elicits inward I_{Ca} precedes a larger depolarizing voltage step. Such prepulses speed up the time course of $I_{K(Ca)}$ activation and thus give rise to significantly increased outward currents with subsequent depolarizations. The extent of this effect depends upon the height and duration of the prepulse. However, the steady-state activation of $g_{K(Ca)}$ never exceeds the value obtained by smaller, single-step depolarizations to the same level of V_m that is reached by the prepulse. Prepulses to between $+10$ mV and $+30$ mV produce the strongest activation, and the optimal prepulse duration was typically between 10 msec and 30 msec. Since the steady-state outward currents grew with increased potentials, the normally bell-shaped isochronal $I_{K(Ca)}$ curves measured between 30 msec and 400 msec became straightened to monotonically increasing curves.

These results can be explained by the following hypotheses: Ca^{++} that enters during the prepulse has, by some saturable reaction, a permissive effect on the opening of the $g_{K(Ca)}$ channels. This saturable reaction causes a transition of the channels into an activated state. Once in the activated state, the $g_{K(Ca)}$ channels can be triggered to open by a subsequent depolarization. This triggering depolarization is provided by the second pulse. According to this hypothesis, it is assumed that the ultimate steady-state value of $g_{K(Ca)}$ is determined not by the amplitude of the second pulse, but rather by the energy level of the Ca^{++} that had entered during the first pulse.

By subtracting the $I_{K(Ca)}$ obtained without prepulse from the $I_{K(Ca)}$ obtained with prepulse, a difference current, I_e, was obtained whose time course reflects the turning on of $I_{K(Ca)}$ subsequent to activation by the prepulse. The time constant τ_0 for the development of I_e ranged between 9 msec and 15 msec in various cells. Significantly, there was no dependence of τ_0 on voltage. Moreover, τ_0 derived from prepulse experiments was similar to the time constant of tail currents (I_t) after correction for contamination by residual I_{Ca}.

IR Pulses

A more direct insight into the activation process of $I_{K(Ca)}$ was achieved by the following approach. $I_{K(Ca)}$ shows very delayed activation with depolarization to near the supposed E_{Ca} (+100 mV to +130 mV), where little I_{Ca} could flow even if Ca^{++} channels were open. A short-lasting repolarizing voltage step would immediately drive the inward I_{Ca} through all available Ca^{++} channels. Proportionality between I_{Ca} and the size of the electric field can be expected in this case since the g_{Ca} is kept constant. By clamping back to the original E_{Ca}, the largely isolated $I_{K(Ca)}$ becomes observable, because I_{Ca} would immediately cease to flow.

Using this procedure, the relationship between the IR potential and the steady-state conductance, $g_{K(Ca)}$, was found to be sigmoidal (Fig. 3). The strongest change of $g_{K(Ca)}$ was produced by varying the IR potential between +40 mV and +70 mV. Making the IR potentials more negative than about +10 mV produced no further enlargements of $g_{K(Ca)}$. This contrasted with the considerable growth of the inward I_{Ca} during the IRs, which showed a more than twofold increase between +10 mV and −100 mV. Minimal aftereffects of IRs were observed between +90 mV and +220 mV.

To account for the striking properties of the relation between the V_m of the initiating IR and the resulting $g_{K(Ca)}$, an attempt was made to fit the curves by assuming that the ratio between activated and available states is determined by an energy distribution function, namely:

$$g_v/(g_{max} - g_v) = \exp[(\alpha - \beta V_m)/12.5 \text{ mV}] \tag{4}$$

where g_{max} is equal to the maximum conductance obtained and g_v is the conductance at given IR voltages of 0.5-msec duration. Half-activation and steepness of the relationship are described by α and β, respectively. With a nonlinear least-squares-fit method for two variables, β was calculated to be close to 1, whereas α, which is analogous to an optimum energy level, was near +50 mV. The exponential expression denotes the energy level of Ca^{++} moving through the potential field of the membrane and is simply taken to describe rates of activation of $g_{K(Ca)}$ at varied voltages. If it is considered that the inverse of the function describing rates of activation (k) applies to activation time constants, equation 4 can be transformed to an equation relating τ_v to τ_0:

$$g_v/g_{max} = k_v/k_{max} = \tau_0/\tau_v \tag{5}$$

and

$$\tau_v = \tau_0\{1 + \exp[(-\alpha + \beta V_m)/12.5 \text{ mV}]\}. \tag{6}$$

Thus, interpretation of the conductance data obtained by IR experiments yields an equation similar to equation 3 and permits calculation of the α and β constants for that equation. The theoretical $I_{K(Ca)}$ curves from equations 1, 2, and 3, using τ_0 derived from the prepulse experiments and the α and β constants calculated from the IR experiments, were compared to the experimental $I_{K(Ca)}$ curves (Fig. 2A). The result is a remarkable fit between the theoretical equations and experimentally determined values.

These results show that the dependence of the response of $g_{K(Ca)}$ on the applied field is well described by a simple application of the Boltzmann energy distribution, which relates electrical field strength to the energy level of Ca^{++} moving in the field. In this view, the opening of the $g_{K(Ca)}$ channel requires the presence of an optimum energy condition. With a surplus of ions having energy levels beyond the minimum activation energy, the reaction at a given site would cease to grow. With a higher field at a more depolarized V_m, the condition would not be met at many sites, and its probability decreases further with further positive increases of the field. An excess of Ca^{++} with kinetic energies below the required value could be present in all situations.

SUMMARY

• A particular advantage of the present formulation is that it can be applied in a direct way to define the time parameters of activation of $I_{K(Ca)}$. The activation time course of $I_{K(Ca)}$ is represented by a term consisting of two factors. One factor defines the actual voltage-dependent rate of transformation of the $g_{K(Ca)}$ channels into a primed state, from which they can be triggered to open by a subsequent depolarization. This factor was determined in IR experiments. The second factor determining the time course of $I_{K(Ca)}$ is a singular, potential-invariant time constant, which limits the time course of the fully activated response. This second time constant was measured in prepulse and I_t experiments.

The results of these experiments suggest that opening of the $g_{K(Ca)}$ channels is a two-step process: A priming reaction, dependent on Ca^{++} entry, has a permissive effect on the channel that allows it to be opened by a subsequent depolarization. The saturable priming step is assumed to depend not on [Ca^{++}]$_i$, but rather on the energy level of Ca^{++} that enters the cell. The energy distribution of these ions is, in turn, dependent on V_m. The priming process itself appears first as a nonconductive state, which slowly returns to rest in the absence of depolarization. Thus, priming may be envisaged as a reversible and electrically silent reaction of Ca^{++} with channel sites. Although depolarization appears necessary to trigger the opening of previously primed channels, the actual kinetics of this depolarization-triggered component of the opening process appear to be voltage-independent. These properties of the $I_{K(Ca)}$ elicited by membrane depolarization must be contrasted with the strictly voltage-independent property of the K$^+$ channel activated in Helix neurons by direct intracellular injection of Ca^{++}.

REFERENCES

Ahmed, Z. and J.A. Connor. 1979. Measurement of calcium influx under voltage clamp in molluscan neurones using the metallochromic dye arsenazo III. J. Physiol. (Lond.) 286: 61.

Akaike, N., K.S. Lee, and A. Brown. 1978. The calcium current of Helix neuron. J. Gen. Physiol. 71: 509.

Eckert, R. and D. Tillotson. 1978. Potassium activation associated with intraneuronal free calcium. Science 200: 437.

Eckert, R., D. Tillotson, and E.B. Ridgway. 1977. Voltage-dependent facilitation of Ca^{2+} entry in voltage-clamped, aequorin-injected molluscan neurons. Proc. Natl. Acad. Sci. 74: 1748.

Gorman, A.L.F. and M.V. Thomas. 1978. Changes in the intracellular concentration of free calcium ions in a pace-maker neurone, measured with the metallochromic indicator dye arsenazo III. J. Physiol. (Lond.) 275: 357.

Heinemann, U., H.D. Lux, and M.J. Gutnick. 1977. Extracellular free calcium and potassium during activity in the cerebral cortex of the cat. Exp. Brain Res. 27: 237.

Heyer, C.B. and H.D. Lux. 1976a. Properties of a facilitating calcium current in pacemaker neurones of the snail, Helix pomatia. J. Physiol. (Lond.) 262: 319.

———— 1976b. Control of the delayed outward potassium currents in bursting pace-maker neurones of the snail, Helix pomatia. J. Physiol. (Lond.) 262: 349.

Hodgkin, A.L. and A.F. Huxley. 1952. A quantitative description of membrane current and its application to conduction and excitation in nerve. J. Physiol. (Lond.) 117: 500.

t#

# # #

# #

# #

Hofmeier, G. and H.D. Lux. 1980. Three distinct effects mediated by calcium ions on electrical membrane properties of *Helix* neurons. In *Proceedings of the XXVIII International Congress of Physiological Science,* Budapest, Akademiai Kiado. (In press.)

Kostyuk, P.G. and O.A. Krishtal. 1977. Effects of calcium and calcium-chelating agents on the inward and outward current in the membrane of mollusc neurones. *J. Physiol. (Lond.)* **270:** 569.

Llinás, R., J.R. Blinks, and C. Nicholson. 1972. Ca-transient in presynaptic terminal of squid giant synapse: Detection with aequorin. *Science* **176:** 1127.

Lux, H.D. and C.B. Heyer. 1977. An aequorin study of a facilitating calcium current in bursting pacemaker neurons of *Helix*. *Neuroscience* **2:** 585.

Lux, H.D. and E. Neher. 1973. The equilibration time course of $[K^+]_o$ in cat cortex. *Exp. Brain Res.* **17:** 190.

Meech, R.W. 1972. Intracellular calcium injection causes increased potassium conductance in *Aplysia* nerve cells. *Comp. Biochem. Physiol.* **42:** 493.

―――. 1974. The sensitivity of *Helix aspersa* neurones to injected calcium ions. *J. Physiol. (Lond.)* **237:** 259.

Meech, R.W. and N.B. Standen. 1975. Potassium activation in *Helix aspersa* neurones under voltage-clamp: A component mediated by calcium influx. *J. Physiol. (Lond.)* **249:** 211.

Inactivation of Delayed K⁺ Current

Based on a presentation by

STUART THOMPSON*

**Department of Biology
and Hopkins Marine Station
Stanford University
Stanford, California 94305**

• In their original description of the membrane properties of the squid giant axon, Hodgkin and Huxley (1952) described two types of time- and voltage-dependent membrane currents that contribute to the initiation and termination of the action potential. The inward I_{Na}, which contributes to action potential initiation, is activated rapidly by depolarization and then inactivates spontaneously. In contrast, the delayed K⁺ current $I_{K(V)}$ shows a slower activation time course but no inactivation, at least within the short time frame Hodgkin and Huxley utilized. Over the past several years, however, it has become clear that $I_{K(V)}$, like I_{Na}, exhibits a time-dependent inactivation.

The results of the work of Thompson et al. on the analysis of $I_{K(V)}$ inactivation in isolated somata of the molluscs *Archidoris* and *Anisodoris* is reviewed here. First, a brief review of the voltage-clamp analysis is presented, and then a quantitative model that describes the data is detailed. Finally, the physiological consequences and possible role of $I_{K(V)}$ inactivation are discussed. (For more extensive discussions, see Aldrich [1979] and Aldrich et al. [1979a,b].)

INACTIVATION OF $I_{K(V)}$ IS DUE TO INTRINSIC CHANNEL PROPERTIES

• Figure 1A illustrates the basic properties of $I_{K(V)}$ inactivation. With the application of a voltage-clamp command, there is an initial large

---

*Work presented here was done in collaboration with R.W. Aldrich, Jr. and P.A. Getting.

The report of this presentation was prepared by G. Yellen.

outward current, which then slowly declines over a several-second period. The right panel of Figure 1A illustrates that repetitive pulses produce a similar reduction in outward current. The obvious question here is whether the decline of the outward current is due to an intrinsic closing of the $I_{K(V)}$ channel or whether it is due to other mechanisms, such as decreased driving force due to redistribution of ion gradients, facilitation or growth of a slow inward current, or inactivation of other outward currents such as I_A (see Connor, this volume) or $I_{K(Ca)}$ (see Meech; Lux; both this volume).

That I_A is involved seems unlikely since cells were selected that had I_A inactivated at -40 mV (the V_h used in these experiments). Figure 1B illustrates the results of one experiment which indicates that the contribution from $I_{K(Ca)}$ or an increased I_{Ca} is minimal. When the cell is bathed in a Ca^{++}-free, Co^{++}-containing medium that blocks I_{Ca}, inactivation of the outward current is unaltered. Inactivation can be greatly reduced, however, by adding TEA to the bath. TEA blocks all outward currents to some degree but is most selective for $I_{K(V)}$. These results are illustrated in Figure 1C.

The progressive reduction in $I_{K(V)}$ is not due to K⁺ accumulation

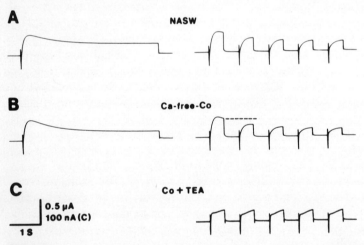

Figure 1
Separation and characterization of $I_{K(V)}$. (A) I_m for an identified pedal ganglion cell 5 or 6 in artificial seawater for a 4.5-sec clamp step to $+10$ mV (*left*) and for repetitive 500-msec pulses to $+10$ mV at a frequency of 1 Hz (*right*). V_h is -40 mV. (B) The same pulse paradigms were presented with the cell bathed in 0 Ca^{++}, 10 mM Co^{++} seawater. (C) The repetitive pulse paradigm was performed on a cell bathed in Co^{++} + TEA saline (0 Ca^{++}, 10 mM Co^{++}, 100 mM TEA replacing an equal amount of Na⁺). Ninety percent of the outward current is blocked; the remaining current is shown at higher gain.

in a restricted space on the outside surface of the cell. An increase in $[K^+]_e$ could affect the currents by moving the E_{rev} more positive, but E_{rev} (measured by tail currents $[I_t]$) is not significantly changed during the pulse trains.

These results, as well as the experiments represented in Figure 1, indicate that the decline of the outward current with a sustained depolarizing command is due to an intrinsic property of the $I_{K(V)}$ channel.

CUMULATIVE INACTIVATION OF $I_{K(V)}$

• $I_{K(V)}$ inactivates during single, long pulses, but during trains of pulses a cumulative inactivation is also seen that is not simply the accumulation of the inactivation appearing during each pulse. Figure 1 illustrates this phenomenon and compares the two methods of examining inactivation. With the pulse trains, the peak current during the second pulse never reaches the final level of the preceding pulse (dashed line, Fig. 1B). Further inactivation occurs between the end of one pulse and the peak of the next. It might seem that inactivation continues during the interval, but, in fact, the extra inactivation occurs at the very beginning of the second pulse, as it reaches its peak (see below).

To explain the cumulative inactivation of $I_{K(V)}$, its voltage dependence and kinetics were studied. The standard measure of cumulative inactivation used was paired pulses, comparing the peak height of the second pulse to the peak height of an identical control pulse not preceded by another pulse. Measured in this way, inactivation depends on the voltage of the first pulse: The amount of inactivation increases and then decreases with increasing depolarizations. Recovery from cumulative inactivation can be measured by varying the interval between the two pulses (see Fig. 5 and below). This recovery is quite slow ($\tau \sim 28$ sec at -40 mV) and is voltage-dependent; the rate of recovery increases markedly at hyperpolarized potentials.

Another technique for looking at the onset of inactivation is illustrated in Figure 2. A fixed test pulse is used to measure the inactivation that occurs during a prepulse of variable duration. The onset of inactivation during the prepulse has a time course that can be fitted by the sum of two exponentials. The fast time constant is similar to that for the turn-on or activation of $I_{K(V)}$; the second matches the time constant for inactivation during a single, long-duration pulse.

KINETIC MODEL FOR $I_{K(V)}$ INACTIVATION

• Thompson et al. have developed a kinetic model for the inactivation of $I_{K(V)}$ based on the idea that its biphasic time course reflects two

distinct modes of inactivation. The only qualitative difference between this model and the Hodgkin-Huxley model for inactivation of
the Na^+ channel is that in this model activation and inactivation are
not independent. In other words, the inactivation process depends on
the activation state (open or closed) of the channel. A kinetic scheme
incorporating such a state-dependent inactivation is shown in the
inset of Figure 3.

Specifically, the model postulates two modes of inactivation—
fast inactivation, which occurs from the closed state, and slow inactivation, which occurs from the open state. Both types of inactivation
are voltage-dependent. Fast inactivation occurs with roughly the
same time course as channel opening and results in a reduction in
peak current. Slow inactivation is the decline in current after the
peak that is seen during single, long pulses. Recovery from both types
of inactivation is slow and voltage-dependent.

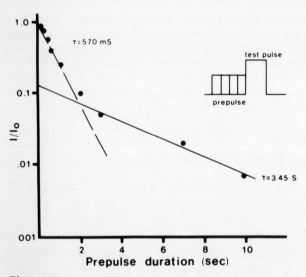

Figure 2
Onset time course for inactivation as determined by a
prepulse method. The pulse paradigm consisted of
prepulses of various durations to 0 mV followed
immediately by a 6-sec test pulse to +20 mV (inset).
The ratio of I/I_0, where I is $I_{peak} - I_\infty$ for the test
pulse with a prepulse and I_0 is $I_{peak} - I_\infty$ for the test
pulse in the absence of a prepulse, is plotted as a
function of prepulse duration. The fast component of
inactivation accounts for approximately 90% of the
total inactivation at 0 mV. The bathing solution was
Ca^{++}-free Co^{++} saline.

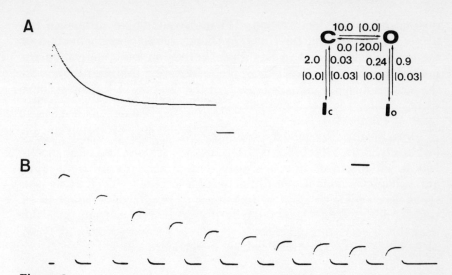

Figure 3
Results of computer simulations using the four-state model for the K^+ channel. (A) Simulated current recorded during a 15-sec pulse from -40 mV to
$+20$ mV. The current rises to an early peak and inactivates to a nonzero
steady state. (B) Simulated currents recorded during 500-msec pulses repeated once per sec. The features of cumulative inactivation are present. Time
scale in A is four times slower than in B. (Inset) The four-state model and the
rate constants used in the simulation. The values in parentheses are those at
the V_h; those not in parentheses are at a V_m of $+20$ mV.

Thompson et al. used their experimental data for the voltage
dependence and kinetics of $I_{K(V)}$ to construct a computer simulation of
the model's predictions for voltage-clamp currents. The rate constants
used in the simulation are listed in the kinetic scheme of Figure 3.
The model successfully reproduces both inactivation during a single
pulse (Fig. 3A) and accumulation of inactivation during a series of
pulses (Fig. 3B).

Figure 4 illustrates how the model produces cumulative inactivation. Two voltage-clamp pulses to $+20$ mV are simulated; the total
current and the occupancies of the four states of the K^+-channel
model are shown. During the rise time of the currents in the first
pulse, two processes compete for the population of closed channels:
$C \rightarrow O$, the opening process, and $C \rightarrow I_c$, the fast inactivation process.
The channels captured by the inactivation processes return only
slowly to the pool of closed channels. Thus, fewer channels are in the
closed state, available for activation, when the second pulse is generated. During the second pulse, the two fast processes once again
compete for this smaller population of closed channels. As a result,

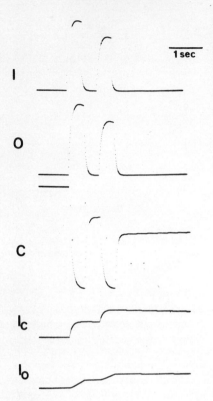

Figure 4
Relative occupancy of the four states of the model during paired pulses. (*Top trace*) Simulated current. Total I_K (I); the population of open channels (O); the population of closed channels (C); the population of inactivated channels due to the fast process (I_c); and the population of inactivated channels due to the slow process (I_o).

the second peak is smaller than the first, and more of the channels are trapped temporarily in the inactivated states. (Notice the accumulation of channels inactivated due to the fast and slow processes and the reduction of closed channels.)

Figure 4 also demonstrates how the model can account for the fact that two short pulses of identical duration, separated by a brief repolarization, produce more inactivation than does a single longer pulse of twice the duration of one of the short pulses (Fig. 1B). Because of the fast $C \rightarrow I_c$ process, inactivation can occur more quickly if the channels are returned momentarily to the closed state by repolarization.

The model also accounts for a peculiar feature of the time course of recovery of $I_{K(V)}$ from inactivation. Recovery is measured by varying the time between two pulses and examining the extent of $I_{K(V)}$ inactivation during the second pulse. At interpulse intervals shorter than 1 sec, inactivation increases with the interpulse interval (Fig. 5A). The time course of this increase is the same as that of the time course of decay of the $I_{K(V)}$ I_ts seen with the closing of $g_{K(V)}$ channels upon repolarization. Fast inactivation occurs only when the channels are

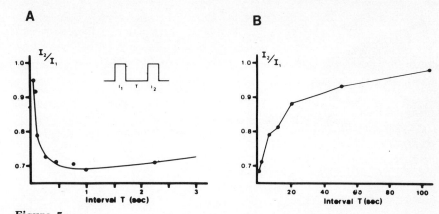

Figure 5
Inactivation recovery time course at −40 mV. Recovery was determined by measuring the ratio of peak outward current (I_2/I_1) during two standard pulses to +10 mV separated by a variable interval T (inset). (A) The ratio of (I_2/I_1) for one typical cell is shown for interpulse intervals below 3 sec. (B) The ratio of I_2/I_1 for the same cell is shown for longer intervals. The overall recovery curve is U-shaped, with a minimum value between 1−2 sec. The recovery curve for times greater than 2 sec (B) can be fit well by a single exponential with time constants averaging 28 sec for four cells. The bathing solution was Ca^{++}-free Co^{++} saline.

closed. Channels that are not yet closed are protected from inactivation. Therefore, $g_{K(V)}$ channels activated by a depolarization pulse that follows an earlier pulse with a very brief delay are less susceptible to fast inactivation than would be the case were they activated after a larger delay. This observation is evidence for a state-dependent model.

Another observation is that elevated $[K^+]_e$ changes the characteristics of use-dependent inactivation, as indicated by a shorter recovery time from inactivation. This can best be described, in terms of the model, as due to an increase in the rate of recovery from the state of inactivation due to fast processes. Possible mechanisms could include the knocking out of a blocking particle (see Armstrong 1971) or a mechanism like that proposed for the inward rectifier (Ciani et al. 1978). However, there is currently no evidence to distinguish among possible mechanisms.

ROLE OF $I_{K(V)}$ INACTIVATION IN SPIKE BROADENING

● In some molluscan neurons, the duration of the action potential becomes progressively longer as action potentials are fired repetitively (Fig. 6). This observation led to the investigation of the possible role of $I_{K(V)}$ inactivation in spike broadening.

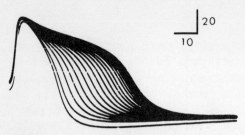

Figure 6
Frequency-dependent spike broadening in
a dorid neuron. Spikes were elicited at a
frequency of 1 Hz and are shown superimposed. 20 mV; 10 msec.

It was found that spikes in neurons showing this phenomenon
are characterized by a prominent shoulder on the falling phase of the
somatic action potential. Spike broadening occurs by accentuation of
this shoulder into a plateau. Both broadening and the action potential
shoulder depend upon the presence of inward I_{Ca}, but no evidence
was obtained for facilitation of g_{Ca} during repetitive voltage-clamp
pulses applied at frequencies at which spike broadening occurs. All
cells studied, however, showed inactivation of outward $I_{K(V)}$ regard-
less of whether or not the somatic action potential broadened during
repetitive firing. In the preceding sections, it was shown that the
decrease in $I_{K(V)}$ during repetitive depolarizations is due to cumulative
inactivation of the voltage-dependent outward current $I_{K(V)}$ (see also
Aldrich et al. 1979b). Spike broadening appears to require the presence of an inward I_{Ca} (see Hagiwara, this volume) and cumulative
inactivation of outward $I_{K(V)}$.

An extension of this work would be to combine the model for
$I_{K(V)}$ inactivation described here with empirical formulae for other
membrane currents. A computer simulation could then be made to
assess quantitatively the role of various membrane currents and channel properties in frequency-dependent spike broadening.

What relevance could frequency-dependent spike broadening due
to the inactivation of $I_{K(V)}$ have for the functioning nervous system? At
this time, it is possible only to speculate. If spike broadening occurs
in nerve terminals (where there is certainly I_{Ca} to maintain the spike
plateau), the increased Ca^{++} influx during the spike might increase
the amount of transmitter release. This suggests one possible mechanism for frequency-dependent homosynaptic facilitation. Another interesting possibility is suggested by the fact that recovery from I_K
inactivation is voltage-dependent. A well-timed IPSP could facilitate
quick recovery from inactivation, which would reset any spike broadening that had occurred and make the spikes narrow again. This is

one mechanism by which synaptic potentials could have effects other than simple summation.

All of the molluscan neurons examined by Thompson et al. show inactivation of $I_{K(V)}$. However, even though this inactivation underlies frequency-dependent spike broadening, not all cells show spike broadening. Thus, although the currents in different cells may be qualitatively similar, the proportions in which the currents are combined can lead to different patterns of excitability in different cells. In this case, a maintained inward current (usually I_{Ca}) is required for the expression of $I_{K(V)}$ inactivation as spike broadening. This inward current maintains the plateau of the action potential until it is counterbalanced by $I_{K(V)}$. Without a maintained inward current, the action potential falls quickly with inactivation of I_{Na}. It should be fascinating to study whether other differences in excitability among cells are based on underlying qualitative differences in membrane currents or, as in this case, are based simply on the proportions of the various currents.

REFERENCES

Aldrich, R.W., Jr. 1979. "Cumulative inactivation of outward current in molluscan neurons and its role in use-dependent broadening of action potentials." Ph.D. dissertation, Stanford University.

Aldrich, R.W., Jr., P.A. Getting, and S.H. Thompson. 1979a. Inactivation of delayed outward current in molluscan neurone somata. *J. Physiol. (Lond.)* **291:** 507.

――――――. 1979b. Mechanism of frequency-dependent broadening of molluscan neurone soma spikes. *J. Physiol. (Lond.)* **291:** 531.

Armstrong, C.M. 1971. Interactions of tetraethylammonium derivatives with the potassium channels of giant axons. *J. Gen. Physiol.* **58:** 413.

Ciani, S., S. Krasne, S. Miyazaki, and S. Hagiwara. 1978. A model for anomalous rectification: Electrochemical potential dependent gating of membrane channels. *J. Membr. Biol.* **44:** 103.

Hagiwara, S., K. Kusano, and N. Saito. 1961. Membrane changes of *Onchidium* nerve cell in potassium-rich media. *J. Physiol. (Lond.)* **155:** 470.

Hodgkin, A.L. and A.F. Huxley. 1952. A quantitative description of membrane current and its application to conduction and excitation in nerve. *J. Physiol. (Lond.)* **117:** 500.

The Fast K$^+$ Channel and Repetitive Firing

Based on a presentation by

JOHN A. CONNOR

Department of Physiology and Biophysics
University of Illinois
Urbana, Illinois 61801

• A transient K$^+$ current quite different from the familiar delayed K$^+$ current, $I_{K(V)}$, described by Hodgkin and Huxley (1952) was discovered by Hagiwara et al. (1961) in the mollusc Onchidium. Since then, this distinctive current—the A current (I_A)—has been demonstrated in many other molluscs (e.g., Connor and Stevens 1971; Gola and Romey 1971; Neher 1971), in arthropods (Connor 1975; Lisman et al. 1979), and in vertebrates (Nakajima 1966). In nearly every case, I_A has been found in cells of the encoder type, that is, neurons that transpose stimulus voltage into repetitive spike activity. Encoder cells are distinguished by a linear relationship between input intensity and output spike frequency. As will be described below, I_A plays an important role in determining the repetitive firing properties of these cells.

TIME- AND VOLTAGE-DEPENDENT PROPERTIES OF I_A

• The time courses of I_A under voltage clamp in a molluscan neuronal cell body (from the marine slug Anisodoris nobilis) and in a crustacean axon (from the crab Callinectes sapidus) are illustrated in Figure 1. Although the kinetics are faster in the axon, in both preparations this current is activated rapidly and then decays with an exponential time course. In each case, the g_A begins to be activated at voltages considerably more negative than spike threshold, at levels of V_m where other voltage-dependent conductances are not turned on. At less negative potentials, fast I_{Na} and $I_{K(V)}$ are activated, which obscures

The report of this presentation was prepared by E.T. Walters.

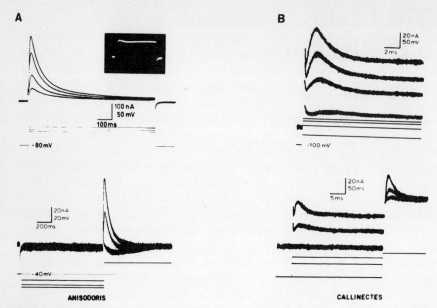

Figure 1
Time course of I_A in a molluscan neuronal cell body (A) and in a crustacean axon (B). (A, inset) Current flow for a large positive step that also activates inward current and $I_{K(V)}$.

the expression of I_A (Fig. 1A, inset). A second property of I_A, which is seen in both preparations, is the requirement for a conditioning hyperpolarization before the activating step for I_A to be activated. This is seen in the lower panels of Figure 1, A and B, where voltage has been stepped from various conditioning levels to a constant level. As the conditioning pulse is made more negative, there is a progressive increase in the magnitude of I_A, even though the test level is the same. Thus, I_A depends upon both the conditioning and test voltages, and in this dual voltage dependence I_A resembles fast I_{Na} (Hodgkin and Huxley 1952).

QUANTITATIVE DESCRIPTION OF I_A

• The similarity of the kinetics of I_A and I_{Na} allows the convenient expression of I_A by means of functions similar to those developed by Hodgkin and Huxley for I_{Na}. Thus, the expression for the current resulting from a positive step from a large negative holding voltage (V_h) to a test voltage (V_T) is

$$I_A = \overline{g}_A \, [A(V_T) \, (1 - \exp^{-t/\tau_A})]^n \, [B(V_h) \exp^{-t/\tau_B}(V_T - E_K)] \qquad (1)$$

$A(V_T)$ is a voltage-dependent activation function (analogous to m_∞ in the Hodgkin-Huxley equations) and is determined from the relationship between V_T and I illustrated in the top panels of Figure 1, A and B. $B(V_h)$ is an inactivation function (analogous to h_∞) determined from the relationship between conditioning voltage and I following the step to the test voltage (Fig. 1, A and B, lower panels). The activation time constant (τ_A) is measured from the rising time course of I_A. The inactivation time constant (τ_B) is determined by the decay of I_A or by the rate at which conditioning voltages reactivate or inactivate I_A. The parameter n is necessary to generate the sigmoidal activation time course of I_A from an exponential fitting function. Fitting procedures (see Connor and Stevens 1971; Connor et al. 1977) have yielded a value for n of 4 in molluscan neurons and 3 in crustacean axons.

The steady-state activation $A(V_T)$ and inactivation $B(V_h)$ functions describing I_A are similar in the molluscan and crustacean preparations (Fig. 2, A and B), although there are some quantitative differences. The most important difference is that the time course of conductance changes is faster in the crustacean axon than it is in the molluscan cell body. This difference parallels differences between the axon and the cell body found for other voltage-dependent conductances. In addition, τ_A and τ_B are voltage-dependent in the crustacean axon (Fig. 2C), but not in the molluscan cell body. Also, reflecting the general finding that neuronal somata operate at more positive voltages than do axons, the activation and inactivation parameters of the cell body are roughly 20 mV more positive than the corresponding axon parameters (cf. Fig. 2, A and B). Finally, the intersection of the A_∞ and B_∞ curves is higher in the axon preparation than in the molluscan cell-body preparation.

EFFECT OF I_A ON INTERSPIKE INTERVAL

• Because of the position of the I_A A and B parameters on the voltage axis, I_A is very sensitive to spike afterhyperpolarization (Fig. 3A). This sensitivity was illustrated in a molluscan preparation using a double-pulse voltage-clamp procedure (Fig. 3B). Each run began at -40 mV, a level near the cell's V_R and at which I_A is largely inactivated. Membrane voltage was then stepped to a conditioning voltage of -65 mV, a level reached during spike afterhyperpolarization. Finally, the voltage was stepped up to -30 mV. In successive runs the time at -65 mV was increased (allowing the B parameter to increase) until the time was comparable to the duration of the spike afterhyperpolarization. As the time at -65 mV was increased, I_A at -30 mV increased. Thus, I_A, which is inactivated during the spike, is partly restored by the spike afterhyperpolarization. In an unclamped neuron

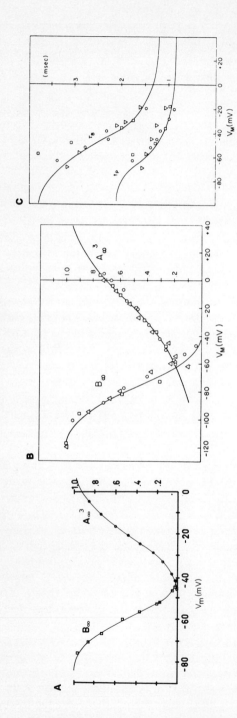

Figure 2

Voltage dependence of I_A conductance parameters in molluscan soma and crustacean axon. (A) A_∞^3 (○) and B_∞ (□) parameters for molluscan (*Anisodoris*) soma. The activation and inactivation time constants, τ_A (3 msec at 18°C) and τ_B (60 msec at 18°C), do not depend upon voltage in this preparation. (B) A_∞^3 (*right*) and B_∞ (*left*) for crustacean axon. (C) Time constants for I_A in crustacean axon. (t_p) Time to peak current (a measure of activation time course).

A

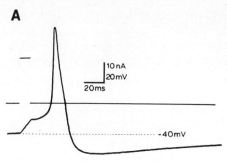

Figure 3
Double-pulse experiment illustrating the recovery of I_A during hyperpolarization. (A) Time course of an action potential showing the spike afterhyperpolarization. (B) I_A in the same cell (on the same time scale) at -30 mV following successively longer hyperpolarizing voltage-clamp steps to -65 mV.

B

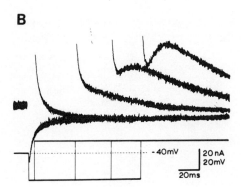

that is given constant current stimulation, the decay of the afterhyperpolarization will cause the A term to increase. Therefore, while both the A and B terms are nonzero, I_A will flow. Since I_A is outward, it will subtract from the stimulus current and thereby prolong the interspike interval (ISI). Because g_A is proportional to the product $A^n B$, the time course of g_A when the voltage is changing is determined by two processes—one that decays and one that grows.

SIMULATION OF REPETITIVE FIRING: THE ROLE OF I_A

• To analyze the effects of the I_A parameters further, it has been useful to make quantitative simulations of repetitive firing. These simulations have shown that the conjoint action of the A and B terms results in a surprisingly slow decay of I_A.

The quantitative simulations shown here utilize crustacean axon data because, first, the measurements in this preparation are uncomplicated by Ca^{++} fluxes or by I_K other than $I_{K(V)}$ known to be present in molluscan cell bodies, and, second, the faster time constants in the axon emphasize differences between the rates of conductance changes under voltage clamp and between the repetitive firing frequencies attainable. Using equations for g_{Na} and $g_{K(V)}$ slightly modified from the Hodgkin-Huxley equations for squid axon, an equivalent circuit of the crustacean membrane with four parallel branches (capacitance, leakage, $g_{K(V)}$, and g_{Na}) was tested for repetitive activity. In response to maintained depolarizations, this model produced repetitive firing that ranged from a minimum of 77 spikes/sec up to a maximum rate of several hundred spikes/sec. Although the maximal firing rate matched that observed in crustacean axons, these axons are actually able to fire spikes at much lower frequencies than the minimum produced by the model—less than 2 spikes/sec compared with 77 spikes/sec.

The performance of the model membrane was greatly improved by adding a fifth parallel branch, g_A, to the circuit. Although retaining the ability to discharge at high frequency, the model's dynamic range was extended downward, reaching a minimum firing frequency of less than 2 spikes/sec, matching the range of crustacean axons (Fig.

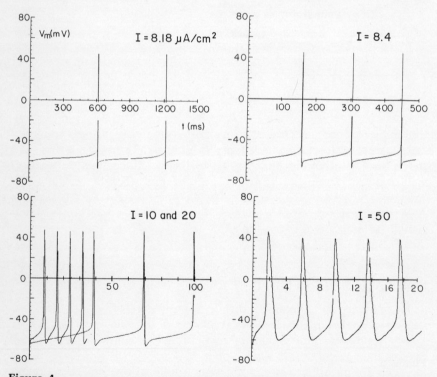

Figure 4
Repetitive firing of a five-branched model of neuronal membrane modified from the Hodgkin-Huxley equations (18°C). This model incorporates I_A in addition to capacitance and the I_{Na}, $I_{K(V)}$, and leakage currents. Values of stimulus current (I) are shown in each panel. (Reprinted, with permission, from Connor et al. 1977.)

4). Examination of this model showed that whereas the voltage trajectory during the ISI is determined by the three dynamic conductance systems (g_{Na}, $g_{K(V)}$, and g_A) and C_m, the controlling factor is g_A (see also Fig. 5A). Following each spike, I_A undergoes a fast, transient increase, followed by a very slow decay. The slow time course of this decay is critical in determining the ISI, and it results from the joint action of the A and B terms in the voltage range traversed during the ISI. Because of this joint action, I_A slowly decreases with depolarization, even though the driving force for I_A ($V_m - E_K$) is increasing. The net result is that the decay of I_A is extremely slow relative to the dynamics of the current under voltage clamp. Thus, I_A can prolong the ISI and allow low-frequency firing.

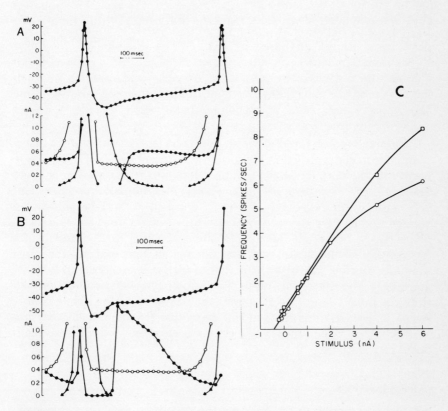

Figure 5
Effects of temperature on I_A and repetitive firing. (A) Computed voltage and current (absolute values) trajectories at 5°C. I_I (o), inward current that is presumably carried by Na; $I_{K(V)}$ (▲); I_A(●). (B) Trajectories at 11°C. I_I (o); $I_{K(V)}$ (▲); I_A (●). (C) Spike frequency vs stimulus intensity computed with parameters measured at warm (□) and cold (o) temperatures. (A and B are reprinted, with permission, from Partridge and Connor 1978.)

TEMPERATURE DEPENDENCE OF I_A AND REPETITIVE FIRING

• If encoder properties are important for neuronal information processing, as seems likely, it would be expected that these properties should be relatively stable in the face of changing organismic conditions. For a cold-blooded animal (such as a marine slug, which moves from deep to shallow water to lay its eggs), changes in temperature inevitably will affect the chemical processes underlying neuronal function. How do animals insure that the critical repetitive firing characteristics of their encoder neurons are not destroyed by changes in temperature?

Partridge and Connor (1978) examined the temperature dependence of I_A in dorid molluscs, which are known to be exposed to rapid changes of temperature in their natural environment. They found that although τ_A and τ_B of I_A are highly temperature-dependent, having a Q_{10} of about 3, spike-frequency versus stimulus-intensity plots were almost unaltered by temperature changes in the lower linear range (where I_A is most important) (Fig. 5C). Voltage-clamp studies showed that two opposing changes occurred in I_A when temperature was increased: Its amplitude increased and its time course decreased. Thus, at higher temperatures, I_A reaches a higher value, but it inactivates more quickly.

Parameters measured at warm and cold temperatures were incorporated into a simplified model of the molluscan cell-body membrane (omitting I_{Ca} and $I_{K(Ca)}$), and simulations of repetitive firing were made at each temperature to evaluate the temperature-dependent effects of I_A on repetitive activity. Figure 5, A and B, shows the voltage and I_m time courses for a stimulus delivered at 5°C and 11°C, respectively. Although there were some changes in $I_{K(V)}$ and in the topography of the spike with the change in temperature, the most dramatic change was in I_A, which showed large increases in both its amplitude and its decay rate. The frequency-intensity plot derived from the model in Figure 5C shows that these opposing changes in I_A result in temperature-independent repetitive firing in the linear frequency range, matching experimental observations of molluscan cell-body preparations.

CONCLUSIONS

• I_A is instrumental in determining the repetitive discharge characteristics of encoder neurons in two preparations, one molluscan and one crustacean. Perhaps the most unique aspect of I_A is that the voltage change it produces under constant current stimulation has a very slow time course relative to the dynamics of I_A under voltage-clamp conditions. Thus, I_A is an effective time-delay mechanism (see Byrne and Koester, this volume). In addition, since relatively little current is required to hyperpolarize neurons to a level at which inactivation of I_A is removed, I_A can provide a simple gating mechanism for turning different responses on and off. For example, a burst of IPSPs could reactivate I_A and greatly delay subsequent spikes. Finally, I_A has built into it a temperature-compensation mechanism that allows the cell to maintain stable repetitive firing characteristics despite the inevitable temperature dependence of the molecular processes underlying its ionic conductances.

REFERENCES

Connor, J.A. 1975. Neural repetitive firing: A comparative study of membrane properties of crustacean walking leg axons. *J. Neurophysiol.* **38:** 922.

Connor, J.A. and C.F. Stevens. 1971. Voltage clamp studies of a transient outward membrane current in gastropod neural somata. *J. Physiol. (Lond.)* **213:** 21.

Connor, J.A., D. Walter, and R. Mckown. 1977. Neural repetitive firing: Modifications of the Hodgkin-Huxley axon suggested by experimental results from crustacean axons. *Biophys. J.* **18:** 81.

Gola, M. and G. Romey. 1971. Responses anomales a des courants sous liminares de certaines membranes somatiques (neurons géants d'*Helix pomatia*). Analyse par la methode du voltage impose. *Pfluegers Archiv. gesamte Physiol. Menschen Tiere* **327:** 105.

Hagiwara, S., K. Kusano, and N. Saito. 1961. Membrane changes of *Onchidium* nerve cell in potassium-rich media. *J. Physiol. (Lond.)* **155:** 470.

Hodgkin, A.L. and A.F. Huxley. 1952. A quantitative description of membrane current and its application to conduction and excitation in nerve. *J. Physiol. (Lond.)* **117:** 500.

Lisman, J.E., M.C. Swan, and G.L. Fain. 1979. Limulus photoreceptors have a transient outward current not dependent on Ca^{2+} entry. *Biophys. J.* **25:** 268a.

Nakajima, S. 1966. Analysis of K inactivation and TEA action in the supramedullary cells of Puffer. *J. Gen. Physiol.* **49:** 629.

Neher, E. 1971. Two fast transient current components during voltage clamp on snail neurons. *J. Gen. Physiol.* **58:** 36.

Partridge, L.D. and J.A. Connor. 1978. Repetitive firing and temperature changes: A mechanism for minimizing temperature effects on the frequency-intensity relation. *Am. J. Physiol.* **234:** C155.

Ionic Conductances in Bursting Pacemaker Cells and Their Hormonal Modulation

Based on a presentation by

THOMAS G. SMITH, JR.*

Laboratory of Neurophysiology
National Institute of Neurological
and Communicative Disorders and Stroke
Bethesda, Maryland 20205

• The types and characteristics of ionic channels in a neuronal membrane determine a number of its electrical characteristics, such as spike threshold, repetitive firing and adaptation properties, and the time course of the action potential. Ionic channels also determine spontaneous activity. For example, not all neurons are electrically silent in the absence of synaptic input. Cells that fire regularly in the absence of any outside influence are called pacemaker cells. Beating pacemaker neurons fire single spikes at regular intervals; bursting pacemaker neurons fire bursts of spikes in a regular pattern during the depolarized phase of the bursting pacemaker potential (BPP) (for a fuller discussion, see Barker and Smith 1978). In this discussion, an attempt is made to characterize the basis for BPP activity in terms of its underlying conductances. The studies concentrate on two identified molluscan bursting neurons—cell R15 in *Aplysia californica* and cell 11 in *Otala lactaea.*

BPP IS DUE TO INTRINSIC CONDUCTANCE MECHANISMS

• Before describing studies on the endogenous membrane properties of these bursting neurons, it is important to consider the evidence that BPP activity in these particular cells is intrinsic to them and is not the result of a recurrent synaptic input. Alving (1968) was the first to show that physically isolated R15 cells continue to have BPPs.

---

*Work presented here was done in collaboration with J.L. Barker.
The report of this presentation was prepared by G. Yellen.

Less dramatic demonstrations that BPP activity is endogenous to these cells are the findings that no oscillatory activity is seen if the cell is hyperpolarized by injecting current or is voltage-clamped to any fixed V_m. Both of these techniques would disrupt feedback mechanisms within the cell but would not directly affect synaptic inputs.

The voltage-clamp steady-state I-V relationship of pacemaker cells, determined by measuring current during long voltage-clamp steps, illustrates why these neurons are not silent (Fig. 1). For all potentials below about -20 mV, net ionic current is inward, and it tends to depolarize the cell. Furthermore, this depolarizing drive is regenerative between about -50 mV and -35 mV; depolarization leads to even more inward current (as indicated by the negative slope of the I-V curve).

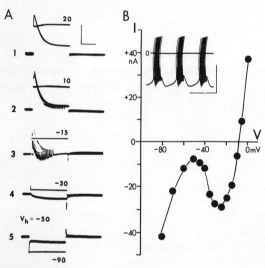

Figure 1
Data from cell R15 *Aplysia californica*. (A) Voltage-clamp data. V (————) and I (▬▬▬) traces for 5-sec pulse commands from V_h of -50 mV to V_m indicated by numbers adjacent to voltage trace. Traces begin with V and I traces superimposed. Positive-going potentials and outward currents are displayed upward. Calibration is 45 mV for voltage traces; 500 nA (1), 200 nA (2), 100 nA (3), and 50 nA (4, 5) for current traces; time calibration is 2.5 sec. (B) Quasisteady-state I-V curve from voltage-clamp data shown in A. (*Inset*) Spontaneous BPP oscillations in the unclamped cell. Calibrations: 50 mV, 26 sec. (Reprinted, with permission, from Smith et al. 1975.)

All pacemaker neurons examined thus far have a persistent inward current, which does not inactivate. This feature contributes to V_m instability; that is, these neurons have no stable V_R (defined as the V_m where net I_m is zero). Whether a pacemaker cell fires in a beating or bursting mode is determined by the nature of the outward currents (see below).

NA$^+$ PACEMAKER CONDUCTANCE

• In *Aplysia* cell R15 and *Otala* cell 11, replacing Na$^+$ with Li$^+$, Tris-H$^+$, or sucrose in the bathing medium abolishes the region of negative slope in the *I-V* curve. The difference between the currents with and without Na$^+$ present is illustrated in Figure 2. This Na$^+$-dependent current is interpreted to be one that is carried by Na$^+$ through a Na$^+$ pacemaker conductance and provides the depolarizing drive for BPP activity in cell R15.

Other investigators have suggested from studies in *Helix* (Eckert and Lux 1976) and in *Tritonia* (Thompson 1976) that the persistent inward current is carried by Ca^{++} (see also Gorman, this volume). In cell R15 and cell 11, reducing [Ca^{++}]$_e$ increases the length and amplitude of the slow BPP and the amount of prolonged inward current seen in voltage-clamp steps (Fig. 3). This result is complicated, however, by the fact that Ca^{++} is known to induce outward currents upon entering the cell (see Meech; Lux; both this volume; see also Fig. 3A), and the disappearance of these outward currents might to some extent offset the disappearance of inward I_{Ca}, if it is present.

As it now stands, proponents of Ca^{++} as the charge carrier must contend with the observation that removing Na$^+$ abolishes the inward current and BPP activity. Proponents of Na$^+$ as the charge carrier

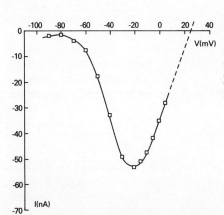

Figure 2
Difference *I-V* curve between *I-V* curves obtained in the presence and in the absence of external Na$^+$ from cell R15 in *Aplysia*. (Reprinted, with permission, from Barker and Smith 1978.)

must explain why Co^{++}, classically a blocker of Ca^{++} channels, also seems to block the slow inward current and BPP activity. It is preferable to think that there is a Co^{++}-blockable g_{Na} present in pacemaker neurons that carries most of the slow inward current.

The turn-on of the Na$^+$ pacemaker conductance is very fast (less than 200 μsec) and shows little inactivation. The fast kinetics of this conductance may have caused other workers (including the proponents of Ca^{++} as the charge carrier) to lump it with the ill-defined instantaneous leakage current.

K$^+$ PACEMAKER CONDUCTANCE

• In contrast to Na$^+$ pacemaker conductance, K$^+$ pacemaker conductance has very slow kinetics and can be seen by analyzing the tail currents (I_t). After the slow g_K is turned on by a depolarizing voltage-clamp step or by BPP activity, the membrane is clamped to a new V_m. The g_K returns only slowly to the steady-state value appropriate to the new potential, and the size of I_t that flows (before the new steady state is achieved) reflects the amount of conductance activated originally.

Figure 4 illustrates the slow K$^+$ I_t after a 5-sec depolarizing voltage-clamp step. The identical clamp sequence was delivered in

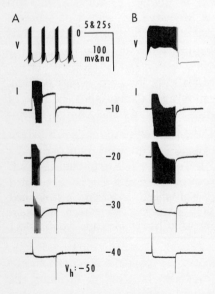

Figure 3
Effect of Ca^{++}-free seawater on BPPs and voltage-clamp currents in cell R15 *Aplysia*. (A) NaCl = 500 mM; CaCl$_2$ = 10 mM; MgCl$_2$ = 60 mM; KCl = 10 mM; Tris-Cl = 15 mM; pH 7.8. (B) Same solution as in A, except CaCl$_2$ = 0 mM. Removal of external CaCl$_2$ increases amplitude and duration of BPP and number of spikes per burst (A V and B V). In A I and B I, V_h = −50 mV. Traces show voltage-clamp currents evoked by 5-sec steps to potentials shown between A and B. Ca^{++}-free solution does not abolish persistent inward current (cf. −40 mV and −30 mV in A and B) but does eliminate most of the late outward current during the step and the slow outward I_t (cf. −10 mV, −20 mV, and −30 mV in A and B). Calibrations: 5 sec and 100 nA for I traces; 25 sec and 100 mV for V traces. (Reprinted, with permission, from Barker and Smith 1978.)

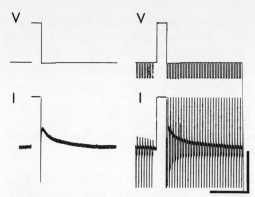

Figure 4
Voltage-clamp I_t from cell R15 in *Aplysia*.
(*Left*) I is I_t when V_m is returned from 5-sec
command to 0 mV (V). (*Right*) I is same as *left*,
plus the currents produced by 0.5-sec, 20-mV
hyperpolarizing pulses before and following
5-sec command to 0 mV (V). Increase in pulse
currents during I_t indicates an increase in g_m
whose time course parallels the decay of I_t.
The large deflections that are superimposed
on the exponential decay of I_t (*right*) are due
to capacitive artifacts. (Reprinted, with permission, from Barker and Smith 1978.)

the right panel, except here the input conductance was measured by
delivering a train of constant, hyperpolarizing voltage commands.
The size of the resultant current pulses is an index of the g_m (from
Ohm's law, $I = gV$). Thus, the right part of Figure 4 illustrates that the
outward I_t is due to an increase in g_m, since larger currents are
produced by the equal potential changes. I_t can be measured at different V_ms; it reverses at a potential determined by the equilibrium
potential of the ion carrying the current. The reversal potential (E_{rev})
of the initial I_t varies with the $[K^+]_e$ in the direction expected if much
of the current is carried by K^+. However, changing the $[K^+]_e$ tenfold
only changes the E_{rev} of the initial I_t by 20 mV, less than one would
expect if the current were carried entirely by K^+. The early I_t presumably is contaminated by other ionic currents also flowing after the
depolarizing step.

Just as I_t can be used to measure K^+ pacemaker conductance
turned on during a clamp step, it can also be used to measure the
size of this conductance in an unclamped membrane. In this way it is
possible to determine how much I_K is present at various times during
the BPP cycle. Figure 5, B and C, shows I_t when the voltage clamp is
turned on at different phases of the BPP cycle. Outward I_ts are

smallest just before the burst of spikes, get larger during the burst, and are largest at the very end of spiking. It appears that the slow turn-on of g_K during the burst of spike activity eventually terminates spiking. As this g_K turns off during the interburst interval, the depolarizing drive of the Na^+ pacemaker conductance dominates, and a new burst of spikes begins.

Ca^{++} INFLUX TURNS ON THE K^+ PACEMAKER CONDUCTANCE

• At least part of the K^+ pacemaker conductance appears to be $I_{K(Ca)}$ (see Gorman; Meech; Lux; all this volume). Stinnakre and Tauc (1973) showed that Ca^{++} enters the cell coincident with the spikes in a burst. This Ca^{++} entry turns on the slow g_K; if Ca^{++} is removed or if D-600 (a Ca^{++} channel blocker) is applied, the late outward currents during a pulse and the slow outward I_ts are abolished (Fig. 3B).

VASOPRESSIN AND EGG-LAYING HORMONE ALTER BURSTING IN CELL 11 IN *OTALA*

• Cell 11 (*Otala*) differs from cell R15 (*Aplysia*) in the seasonal nature of its BPP activity. In the hibernating phase of its life cycle,

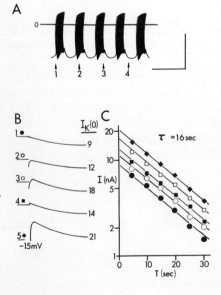

Figure 5
Currents recorded by voltage-clamping at different phases of the spontaneous BPP cycle. (A) Arrows indicate times during the BPP cycle when the cell was clamped to −50 mV. (B, 1-4) Line drawings of the currents recorded at the times numbered in A. Zero current indicated by the horizontal lines below the numbers in each trace. (B, 5) I_t recorded when the cell was returned to V_h (−50 mV) from a 5-sec step to −15 mV. (C) Semilog plot of current time for currents shown in B. Slowly declining currents all fit an exponential with τ of decay = 16 sec. Zero time was the onset of the clamp for 1-4 and the beginning of the I_t in 5. Zero time intercepts of each of these curves are listed under $l_K(0)$ in B. Calibrations: 60 mV, 30 sec in A; 60 nA, 30 sec in B. (Reprinted, with permission, from Smith et al. 1975.)

cell 11 is either silent or generates random or semiregular spikes (Barker and Gainer 1974; Barker et al. 1975; Fig. 6, inset—control). During this time, the zero current point of the voltage-clamp steady-state I-V curve shifts to a more negative value of V_m (Barker and Smith 1976; see also Fig. 6, graph).

One of the more interesting aspects of pacemaker potentials is that they can be modulated or modified by neurohormones. The bag

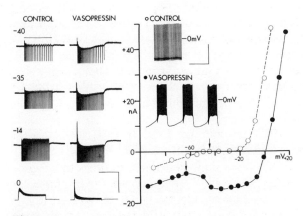

Figure 6
Vasopressin alters voltage-clamp steady-state I-V curve. Recordings are from peptide-sensitive cell 11 from *Otala* before (control) and after (vasopressin) the bath application of 1 μM vasopressin. Firing pattern is illustrated in insets of I-V plot (right). Control trace shows beating pacemaker activity; lower trace shows bursting activity induced by vasopressin. (Left) Membrane of cell voltage-clamped and 5-sec voltage steps imposed (during time indicated by bar, upper left). Currents are shown at different depolarizing voltage steps (to V_ms indicated by numbers above traces) under control conditions and in the presence of vasopressin. Rapid downward-current events represent action potential currents from uncontrolled regions of the cell. Presence of slow inward current that decreases during the command is apparent in the vasopressin-treated membrane. (Right) I-V curve derived from quasi-steady-state currents using most-negative or least-positive current evoked after 1 sec during command. V_h, −45 mV in control and −65 mV in vasopressin (↓). Calibrations: (Left) 10 nA (upper three traces) and 40 nA (lowermost traces), 5 sec. (Right, inset) 50 mV, 20 sec. (Reprinted, with permission, from Barker and Smith 1976.)

cells in *Aplysia* are neurosecretory cells. When stimulated, they release a neuroactive chemical. Instead of being released close to the target cells at specialized synaptic contacts, this "neurohormone" seems to be released directly into the circulation. If a crude extract of bag cells is injected into *Aplysia*, it triggers a set of behaviors that culminates in egg-laying. Chiu et al. (1979) have purified the peptide egg-laying hormone (ELH) that has the same behavioral effect as the crude extract.

Either bag-cell extract (BCE) or antidiuretic hormone (lysine vasopressin [LVP]) can cause the nonbursting cell 11 in hibernating *Otala* to start BPP activity (Fig. 6, inset). Only nanomolar to micromolar amounts of LVP are required for the effect, which develops in seconds to minutes after the peptide is added to the bath. In contrast to this rapid onset, the effect takes hours to reverse after vasopressin is removed. The excitatory effect is specific to cell 11 in *Otala*, but LVP or ELH can also enhance BPP activity in cell R15 in *Aplysia* (Barker and Smith 1978).

The production of BPP activity by LVP or BCE leads to the appearance of a negative-slope region in the *I-V* curve of cell 11 (Fig. 6, graph). As in *Aplysia*, the inward current that produces this region of negative slope is carried by Na^+. Thus, LVP and BCE appear to modulate or activate Na^+ pacemaker conductance, as well as to increase the size of K^+ pacemaker conductance (Barker and Smith 1976, 1978).

SUMMARY

• It is likely that a similar neurohormone, endogenous to *Otala*, is responsible for the seasonal change in the activity of cell 11. The action of vasopressin or the endogenous hormone on other cells has not yet been well-characterized. However, it has been observed in cell 11 and cell R15 that a hormone-induced change displacing the delicate balance of ionic currents in a neuron can produce major changes in firing pattern. The circulatory route of delivery of neurohormones suggests that they may have many target cells. It should be interesting and challenging to find these cells, sort out the various mechanisms of hormone effects on neurons, and discover how these assorted individual effects produce the concerted behavioral changes triggered by the hormone.

REFERENCES

Alving, B.O. 1968. Spontaneous activity in isolated somata of *Aplysia* pacemaker neurons. *J. Gen. Physiol.* **51:** 29.

Barker, J.L. and H. Gainer. 1974. Peptide regulation of bursting pacemaker activity in a molluscan neurosecretory cell. *Science* **184:** 1371.

Barker, J.L. and T.G. Smith, Jr. 1976. Peptide regulation of neuronal membrane properties. *Brain Res.* **103:** 167.

————. 1978. Electrophysiological studies of molluscan neurons generating bursting pacemaker potential activity. In *Abnormal neuronal discharges* (ed. N. Chalazonitis and M. Boisson), p. 359. Raven Press, New York.

Barker, J.L., M. Ifshin, and H. Gainer. 1975. Studies of bursting pacemaker potential activity in molluscan neurons. III. Effects of hormones. *Brain Res.* **84:** 501.

Chiu, A.Y., M.W. Hunkapiller, E. Heller, D.K. Stuart, L.E. Hood, and F. Strumwasser. 1979. Purification and primary structure of the neuropeptide egg-laying hormone of *Aplysia californica*. *Proc. Natl. Acad. Sci.* **76:** 6656.

Eckert, R. and H.D. Lux. 1976. A voltage-sensitive persistent calcium conductance in neuronal somata of *Helix*. *J. Physiol. (Lond.)* **254:** 129.

Smith, T.G., J.L. Barker, and H. Gainer. 1975. Requirements for bursting pacemaker activity in molluscan neurones. *Nature* **253:** 450.

Stinnakre, J. and L. Tauc. 1973. Calcium influx in *Aplysia* neurones detected by injected aequorin. *Nature New Biol.* **242:** 113.

Thompson, S.H. 1976. "Membrane currents underlying bursting in molluscan pacemaker neurons." Ph. D. dissertation, University of Washington, Seattle. University Microfilms International, Ann Arbor, Michigan.

Applicability of Channel Analysis in Molluscs to Vertebrate Central Neurons

Based on a presentation by

RUDOLFO LLINÁS

Department of Physiology & Biophysics
New York University Medical Center
New York, New York 10016

• Since the original demonstration of I_{Na} and I_K underlying the action potential in the giant axon of the squid, molluscan preparations have provided fundamental insights into many of the ionic currents responsible for neuronal activity. Until recently, only a few of the currents examined in molluscs had been described in vertebrate neurons. It now appears, however, that all of the major currents of the molluscan neuronal cell body can be found in vertebrate neurons, including those of mammals, and, indeed, it appears that some mammalian neurons may have, in addition, novel currents not described in invertebrates.

Some of the currents indicated by recent studies of guinea pig central neurons in in vitro preparations are discussed here. These studies have involved three types of neurons—cerebellar Purkinje cells, neurons in the inferior olive, and vagal motoneurons.

PURKINJE CELLS

• The relatively large size (35-μm soma diameter) and characteristic laminar organization of Purkinje cells in the cerebellar cortex have made possible an elucidation of the patterns of electroresponsiveness in different membrane regions of this extensively investigated neuron (see, e.g., Llinás et al. 1968, 1977). However, it is not possible to examine the ionic basis of these properties in sufficient detail because of the difficulty in maintaining intracellular recordings and in modifying the extracellular ionic medium in a precise manner. To examine ionic mechanisms of somatic and dendritic electrophysiological properties, Llinás and Sugimori (1980a), using procedures first developed

The report of this presentation was prepared by E.T. Walters.

for the hippocampus (Yamamoto and McIlwain 1966), prepared thin (200-μm) sagittal slices of the cerebellum that could be maintained in vitro in a small recording chamber. This preparation allowed direct visualization of unstained Purkinje cell somata (and sometimes dendrites) during impalement, recording from individual cells for 2 hours or more, and rapid change of the ionic composition of the perfusing medium as well as application of various drugs.

Properties of the Purkinje Cell Soma

When an electrode was inserted into the soma of a Purkinje cell in vitro, the observed electrical activity was quite similar to that seen in vivo. Antidromic action potentials (initiated by extracellular stimulation of the Purkinje cell axon in the white matter), synaptic potentials (primarily from climbing fiber activation), and direct electroresponsiveness to injected current were demonstrated. The response to both short and long pulses of depolarizing current was repetitive firing, which with larger current injections became marked by oscillations and bursts of very high frequency firing (Fig. 1A). One interesting finding was that despite the apparent uniformity in appearance of Purkinje cells, different Purkinje cells with similar input resistances showed quite different sensitivities to injected current, as judged by the resulting firing frequencies, indicating different integrative properties. However, all cells showed the same basic patterns of electroresponsiveness and the same ionic dependencies. To gain insight into the conductances underlying the electrophysiological properties recorded in the Purkinje cell soma, various currents were eliminated by ion substitution and with pharmacological blocking techniques.

Ca^{++}-dependent Action Potentials in the Absence of I_{Na}

When I_{Na} was blocked by application of TTX or by replacement of extracellular Na^+ with Tris or choline (Fig. 1B), the fast high-frequency spikes were abolished. Those remaining were slow, broad action potentials distinguished by a slow rate of rise, high threshold, and small amplitude. When I_{Ca} as well as I_{Na} was blocked by application of Co^{++}, Cd^{++}, or Mn^{++}, or by the organic blocker D-600, these slow spikes were abolished and all electroresponsiveness was eliminated. As will be discussed below, the slow Ca^{++} spikes have a dendritic origin, whereas the fast Na^+ spikes originate in the soma.

Na^+-dependent Action Potentials in the Absence of I_{Ca}

Blockade of I_{Ca} with Co^{++}, Cd^{++}, Mn^{++}, or D-600 revealed two distinct Na^+-dependent responses. Both can be seen in Figure 1, C and D. Prolonged pulses of depolarizing current produced, after a delay, a

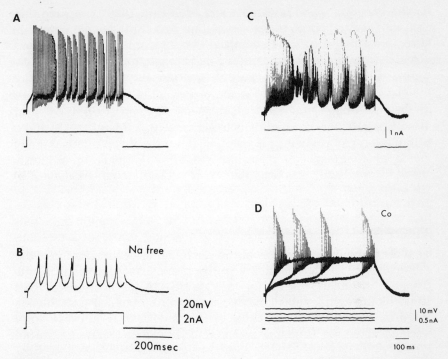

Figure 1
Somatic action potentials. (*A*) Control record showing repetitive firing resulting
from direct stimulation. (*B*) Same cell 30 min after replacing media with Na$^+$-
free solution. (*C*) Control response of another cell before addition of Co^{++}. (*D*)
Response after g_{Ca} was blocked by Co^{++}. As the stimulus was increased, the
latency of onset of firing decreased, but the rate of rise and final plateau
amplitude were independent of stimulus intensity. (Reprinted, with permission, from Llinás and Sugimori 1980a.)

slowly rising response capped with fast action potentials. Fast action
potentials decreased in amplitude progressively as the response stabi-
lized at a plateau depolarization (about −30 mV) until finally there
was complete inactivation of the fast responses. Increasing levels of
depolarizing current decreased the latency of the all-or-none slow
plateau response, but the amplitude of the plateau remained constant
(Fig. 1D). During the plateau potential there was a fivefold decrease
in input resistance. Addition of TTX or replacement of extracellular
Na$^+$ to block I_{Na} abolished both the fast action potentials and the slow
plateau response. These observations indicate that there are two g_{Na}s.
First, there is a typical, fast, inactivating conductance responsible for
the fast spikes. In addition, there is a slow, noninactivating conductance that produces the slow plateau potential. Since the plateau

potential could be shortened by further injection of depolarizing current, the plateau probably represents an equilibrium between the noninactivating g_{Na} and a noninactivating g_K. Support for this interpretation came from intracellular injection of TEA to block $g_{K(V)}$; this caused increased depolarization during the Na$^+$-dependent plateau.

When extracellular Ca^{++} was replaced by Ba^{++}, which is thought to move through the Ca^{++} channel more easily than Ca^{++} itself (Hagiwara 1973) and which fails to activate $g_{K(Ca)}$ (Meech 1978), direct stimulation of Purkinje cells resulted in enormous, prolonged action potentials that reached a plateau of about 55 mV and lasted as long as 10 min. This result suggests that I_{Ca} is probably counterbalanced by an outward $I_{K(Ca)}$.

Properties of the Purkinje Cell Dendrite

One of the major advantages of the cerebellar slice preparation is that it allows precise determination of the microelectrode recording site. This is particularly useful for mapping dendritic electroresponsiveness. Figure 2 illustrates a composite picture (obtained from impalement of different Purkinje cells) of the pattern of responses seen at various distances from the soma. Injection of depolarizing current

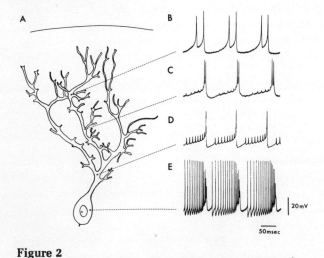

Figure 2
Composite picture of Purkinje cell showing the relationship between somatic and dendritic action potentials during injection of depolarizing currents through the recording electrode. A shift in amplitude of the fast soma spikes as compared with the slow dendritic spikes is seen as the electrode moves from the somatic layer (E) to the superficial layer (B). (Reprinted, with permission, from Llinás and Sugimori 1980b.)

through the recording electrode produced three kinds of action potentials: a fast spike generated by a g_{Na}, a slower Ca^{++}-dependent spike, and a slow Na^+-dependent plateau. The fact that the amplitude of the fast, Na^+-dependent spike was reduced progressively with increasing distance from the soma demonstrated that these spikes are generated in the soma and passively conducted into the dendrites. In addition, the slow, noninactivating Na^+ responses are more prominent in the soma. Conversely, the slow, Ca^{++}-dependent spikes were large in more superficial dendrites and quite small in the soma, indicating that in the dendrites g_m is dominated by g_{Ca}.

Two kinds of Ca^{++}-dependent dendritic electroresponsiveness were found: Ca^{++} plateau potentials and dendritic spike bursts. Both Ca^{++}-dependent responses can be seen in Figure 3. In the presence of TTX, increasingly large depolarizations produced progressively longer plateau potentials (outlasting the stimulus), upon which were superimposed dendritic spike bursts. Addition of Cd^{++} to the bath or elimination of extracellular Ca^{++} abolished both responses. Thus, each response is Ca^{++}-dependent and TTX-insensitive. In addition, both have their lowest threshold and greatest amplitude in the distal dendrites. They differ primarily in the absence of inactivation in the plateau potential, which can last for several hundred msec. The dendritic spike bursts were characterized by their slow onset, large amplitude in the dendrites, multiple spike components (reflecting multiple generation sites), and termination with an afterhyperpolarization that is probably mediated by $g_{K(V)}$ and $g_{K(Ca)}$.

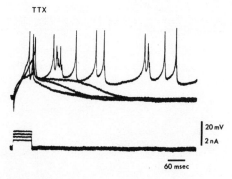

TTX

20 mV

2 nA

60 msec

Figure 3
Ca^{++}-dependent plateau and spike burst seen in Purkinje cell dendrites in the presence of TTX. As the short stimulus is intensified, the local response prolongs until a large plateau potential and superimposed dendritic spike bursts result. (Reprinted, with permission, from Llinás and Sugimori 1980b.)

INFERIOR OLIVARY CELLS

• The inferior olive, a nucleus in the brainstem, is the site of origin of the climbing fiber input to the cerebellar Purkinje cells. Using the same techniques employed in preparing cerebellar slices, Llinás and Yarom (1980) obtained thin sagittal sections of guinea pig brainstem for in vitro examination of inferior olivary neurons. Direct (as well as antidromic) stimulation of these neurons produced three distinct electrophysiological responses: a fast spike, a late depolarization, and a prolonged afterhyperpolarization (Fig. 4A). Blocking g_{Na} with TTX eliminated the initial fast portion of the spike, leaving the late depolarization and afterhyperpolarization intact (Fig. 4B). Conversely, blockers of I_{Ca}, such as Mn^{++}, Cd^{++}, or Co^{++}, abolished the late depolarization and afterhyperpolarization, leaving the fast spike (Fig. 4D). Thus, the fast spike appears to be mediated by Na^+, whereas the late depolarization and afterhyperpolarization appear to be produced by a Ca^{++} spike followed by a $I_{K(Ca)}$. Because this Ca^{++} spike has a high threshold, it is probably generated in the dendrites.

The above conductances are similar to those observed in Purkinje cells. In addition, a novel g_{Ca} was found that is inactivated at resting levels (Fig. 5). In the presence of TTX, when a pulse that is subthres-

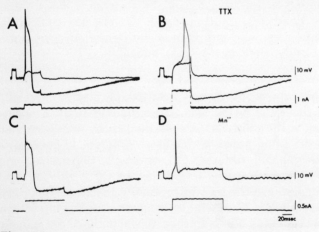

Figure 4
Dependence of inferior olivary cell spikes on Na^+ and Ca^{++}. (A) Control spike and subthreshold membrane response in upper traces (amplitude of current injection in lower trace). (B) Spike and subthreshold response in TTX. Note increased threshold, disappearance of initial portion of the spike, amplitude reduction, and large afterhyperpolarization. (C) Control spike. (D) Same cell in 10 mM $MnCl_2$ to block Ca^{++} and Ca^{++}-activated outward currents. Note disappearance of afterdepolarization and afterhyperpolarization.

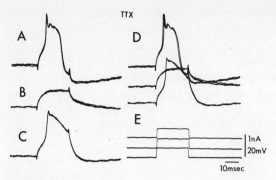

Figure 5
Ca^{++} spikes in inferior olivary cells in TTX.
Traces in A, B, and C are the same as those in D,
but they are separated to show individual responses. In D, a depolarizing stimulus at resting
potential (*middle trace*) produces a subthreshold
response (B). If a steady depolarizing current is
applied first (E, *upper trace*), the previously subthreshold stimulus evokes an action potential (A
and D, *upper record*). If a steady hyperpolarizing
current is applied (E, *lower trace*), the same previously subthreshold stimulus also generates an
action potential (C and D, *lower record*).

hold at resting level was superimposed upon a steady depolarization,
the large Ca^{++} spike shown previously resulted (cf. Figs. 5A and 4B).
However, when the same pulse was superimposed upon a steady
hyperpolarization, a different Ca^{++}-dependent spike was generated
(Fig. 5C). Thus, the same pulse, which is insufficient to generate a
spike at rest, will generate different Ca^{++} spikes at depolarized and
hyperpolarized levels. Both spikes are blocked by Cd^{++} and Co^{++} and
are TTX-resistant. Because the second (normally inactivated) g$_{Ca}$ can
generate a low-threshold Ca^{++} spike by somatic stimulation, it was
concluded that its initiation site is in the cell soma. Since the somatic
Ca^{++} spike inactivates, it exhibits refractoriness when pulses are
paired at short intervals, whereas the noninactivating dendritic Ca^{++}
spikes show no refractoriness. The differences in inactivation are
especially evident when the neurons are bathed in TTX and extracellular Ca^{++} is replaced by Ba^{++}, which fails to activate g$_{K(Ca)}$. Under
these conditions, the spike elicited at a relatively depolarized level
(the dendritic Ca^{++} spike) is greatly prolonged and repolarizes very
gradually. However, the spike elicited following the hyperpolarization (the somatic Ca^{++} spike) is quite brief and repolarizes rapidly as
a consequence of its rapid inactivation.

One of the striking characteristics of inferior olivary cells is that they fire regularly at a rate of about 4 impulses/sec. The electrophysiological properties described above should contribute to such oscillations if V_m is kept relatively hyperpolarized. Thus, a Na^+ spike activates dendritic g_{Ca} and $g_{K(Ca)}$, which produce an afterhyperpolarization that can lead to a rebound spike (generated by the removal of somatic I_{Ca} inactivation). The rebound spike should retrigger the sequence, thereby causing the cell to oscillate (Fig. 6A). In the presence of TTX, the rebound spike is observed easily (Fig. 6B). Moreover, the period of the inherent oscillator can be modulated by the V_m of the cell; hyperpolarization (e.g., produced by current injection or by the drug harmaline, which produces repetitive firing in these cells) increases the firing frequency (Fig. 6B).

Another interesting feature of these cells is that the modulation of Ca^{++}-spike activity by V_m is time-dependent. Steady depolarization produced an immediate reduction in input resistance, which then gradually increased over several seconds. During this time, repeated depolarizing current pulses produced progressively increasing voltage steps until Ca^{++} dendritic spikes were activated. These observations indicate that depolarization activates a $g_{K(V)}$ that slowly inactivates, eventually allowing Ca^{++}-spike generation (Llinás and Yarom 1980).

VAGAL MOTONEURONS

• The brainstem slice preparation also allows convenient access to vagal motoneurons. These cells, in addition to having conductances similar to those described in Purkinje and inferior olivary cells, have an inactivating fast g_K. This conductance was revealed following hyperpolarizations from the resting level. Instead of returning to the resting level with a passive time constant, the potential remained hyperpolarized for approximately 30 msec after the pulse. During this afterhyperpolarization, there was a large conductance increase to ions with an equilibrium potential (measured with biphasic current pulse injections) similar to that of the spike afterhyperpolarization. These properties are similar to those of the fast I_K (I_A) observed in molluscs (see also Connor; Byrne and Koester; both this volume).

CONCLUSIONS

• Because voltage-clamp techniques are difficult to apply to mammalian central neurons, direct measurements of currents in the neurons discussed above have not yet been obtained. However, considerable

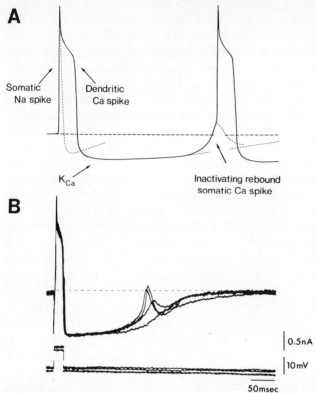

Figure 6

Mechanisms of inferior olivary cell oscillation. (A) Schematic of events generating oscillation. Antidromic or direct
stimulation generates a fast somatic Na+ spike (- - - - -).
This spike triggers a dendritic Ca++ spike, which produces
a long afterdepolarization that in turn triggers a large g_K.
This g_K then generates the prolonged afterhyperpolarization, which removes inactivation from the somatic g_{Ca},
resulting in a rebound depolarization that can retrigger the
entire sequence. (B) Rebound Ca++ spike in the presence of
TTX. (Left) Direct stimulation produces a dendritic Ca++
spike followed by an afterhyperpolarization and rebound
spike. Small changes in steady hyperpolarizations (note
current record) facilitate the rebound spike, making it larger and of shorter latency. (Right) Further hyperpolarization
generates a clear rebound spike and, on one occasion, a
secondary rebound.

information on the ionic conductances underlying the currents has
been gathered using ion substitution and various blocking agents.
These techniques have provided solid evidence in mammalian neu-

rons for all of the conductances that have been studied extensively in molluscs. These include a brief, rapidly inactivating g_{Na} and a late $g_{K(V)}$ like those described by Hodgkin and Huxley (1952) in the squid axon. In addition, there is a slow, noninactivating g_{Na} and a slow, noninactivating g_{Ca}, both of which have similarities to the slow inward currents in molluscan neuronal cell bodies (see Adams et al. 1980; T. Smith, this volume). Mammalian neurons also have a $g_{K(Ca)}$ as well as a fast g_K resembling the conductances underlying $I_{K(Ca)}$ and I_A in molluscs (see Meech; Lux; Connor; all this volume). Finally, there is a novel g_{Ca} in the inferior olivary cell soma that has the unusual property of being inactivated, like the g_A, at V_R.

These findings indicate that the detailed analyses of various ion channels that are emerging from technically advantageous molluscan preparations have direct applicability to the mammalian brain. In addition, recent advances in the biophysics of mammalian neurons suggest that it may soon be possible to analyze in detail both the mammalian currents that are shared with simpler animals and other currents that may be unique to mammals.

REFERENCES

Adams, D.J., S.J. Smith, and S.H. Thompson. 1980. Ionic currents in molluscan soma. *Annu. Rev. Neurosci.* **3**: 141.

Connor, J.A. and C.F. Stevens. 1971. Prediction of repetitive firing behavior from voltage clamp data on an isolated neurone soma. *J. Physiol. (Lond.)* **213**: 31.

Hagiwara, S. 1973. Calcium spikes. *Adv. Biophys.* **4**: 71.

Hagiwara, S., K. Kusano, and N. Saito. 1961. Membrane changes of *Onchidium* nerve cell in potassium-rich media. *J. Physiol. (Lond.)* **155**: 470.

Hodgkin, A.L. and A.F. Huxley. 1952. A quantitative description of membrane current and its application to conduction and excitation in nerve. *J. Physiol. (Lond.)* **117**: 500.

Llinás, R. and M. Sugimori. 1980a. Electrophysiological properties of *in vitro* Purkinje cell somata in mammalian cerebellar slices. *J. Physiol. (Lond.)* **305**: 171.

————. 1980b. Electrophysiological properties of *in vitro* Purkinje cell dendrites in mammalian cerebellar slices. *J. Physiol. (Lond.)* **305**: 197.

Llinás, R. and Y. Yarom. 1980. Electrophysiological properties of mammalian inferior olivary cells *in vitro*. In *The inferior olivary nucleus* (ed. J. Courville et al.), p. 379. Raven Press, New York.

Llinás, R., M. Sugimori, and K. Walton. 1977. Calcium-dependent spikes in the mammalian Purkinje cells. *Neurosci. Abst.* **3**: 58.

Llinás, R., C. Nicholson, J. Freeman, and D. Hillman. 1968. Dendritic spikes and their inhibition in alligator Purkinje cells. *Science* **160**: 1132.

Meech, R.W. 1978. Calcium-dependent potassium activation in nervous tissues. *Annu. Rev. Biophys. Bioeng.* **7**: 1.

Yamamoto, C. and H. McIlwain. 1966. Electrical activities in thin sections from the mammalian brain maintained in chemically defined media *in vitro. J. Neurochem.* **13**: 1333.

Neural Mechanisms Underlying the Stimulus Control of Ink Release in *Aplysia*

Presented by

JOHN H. BYRNE*

Department of Physiology
University of Pittsburgh School of Medicine
Pittsburgh, Pennsylvania 15261

JOHN KOESTER

Division of Neurobiology and Behavior
Department of Physiology
College of Physicians & Surgeons of Columbia University
New York, New York 10032

• For many years, studies of the neural control of behavior focussed primarily on the organization of the nervous system. Neurons were thought of as more-or-less interchangeable units that add up incoming synaptic input until a threshold value is reached, at which point an action potential is triggered. Considerable emphasis was placed on the connections of the single neuron with other neurons, and any specialized electrophysiological properties of the cell received less attention.

More recently, greater efforts have been made to determine the critical cellular features of the individual neuronal types found at the different points in a neural circuit. These studies have revealed an impressive diversity in the functional properties of various neurons. Individual nerve cells differ in their spontaneous firing patterns, as well as in their responses to impinging synaptic input. These variations arise because of cell-to-cell differences in the kinetics, voltage-sensitivities, spatial distributions, and densities of the different synaptic and voltage-sensitive channels that are found in nerve cells. It has been possible with the ink-release system of *Aplysia* to relate the biophysical properties of individual neurons to their specific roles in generating behavior.

INKING BEHAVIOR

• Ink release by *Aplysia* is presumed to serve a defensive function since it occurs only in response to noxious stimuli. Because the stores

*Work presented here was done in collaboration with N. Dieringer and E. Shapiro.

of ink in the ink gland are replenished relatively slowly following release, it would seem advantageous for *Aplysia* to ignore trivial stimuli and to ink only in response to particularly threatening stimuli. But once ink is to be released, a rather massive response might be expected, since the released ink is susceptible to rapid dilution by wave action. This expectation was borne out in behavioral studies in which electrical shocks to the head were used as the noxious stimuli. Ink release was found to have a rather high threshold when measured in terms of stimulus amplitude (Carew and Kandel 1977a) or duration (Fig. 1) (Shapiro et al. 1979). Once threshold is reached, ink release increases steeply for further increases in stimulus duration or intensity. In addition to these static input-output relations, ink release also shows a form of short-term plasticity. The first of a pair of subthreshold noxious stimuli potentiates strongly the effects of the second stimulus, putting it above threshold for triggering inking (Carew and Kandel 1977b).

Our studies on the neuronal basis of the stimulus control of inking have concentrated on the temporal properties of the stimulus-response relations of this behavior. What accounts for the selective insensitivity of ink release to brief stimuli, and what is the mechanism by which the first of two stimuli potentiates the effect of the later stimulus?

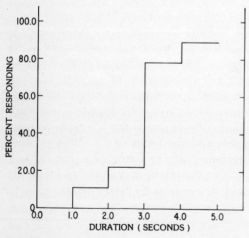

Figure 1
Percentage of animals inking vs stimulus duration. A steep increase in the percentage of animals responding occurred as stimulus duration was increased from 3 sec to 4 sec. (Reprinted, with permission, from Shapiro et al. 1979.)

SELECTIVE INSENSITIVITY OF INK RELEASE
TO BRIEF STIMULI

• A convenient place to begin an analysis of the neuronal properties responsible for the stimulus control of a behavior is at the level of the motor neurons. Carew and Kandel (1977a) have shown that ink release is controlled by a cluster of three, virtually identical, ink gland motor neurons, called the L14 cells. We found that the relatively greater effectiveness of longer-duration stimuli in eliciting ink release could not be explained by temporal facilitation at the level of the motor neuron-effector junction. For a constant-frequency train of action potentials in an ink gland motor neuron, ink release was found to be directly proportional to the duration of the train. Thus, for spike trains up to 5 sec in duration, the first spike was equal in effectiveness to the last spike in producing ink release (Shapiro et al. 1979).

This lack of facilitation, together with the relative ineffectiveness of brief stimuli in triggering ink release, suggest that the initial part of a noxious stimulus should be less effective than the later parts of the stimulus in triggering spikes in the motor neuron. This expectation was confirmed in experiments in which activity of an ink gland motor neuron was recorded during the application of noxious stimuli to the head (Shapiro et al. 1979). Similar results were obtained in later experiments in which the sensory stimulus was mimicked by stimulation of the connectives that carry synaptic input from the head ganglia to the L14 cells in the abdominal ganglion (Byrne et al. 1979). For a 5-sec stimulus, the first $1-3$ sec of stimulation elicited at most a few and sometimes no spikes in the motor neuron. This initial pause in activity was followed by a rapidly accelerating, high-frequency burst of spikes that triggered ink release. Such a delayed response of the motor cell could have two possible explanations: Either the cell has intrinsic properties that cause it to respond sluggishly to excitatory input, or there may be temporal facilitation of the excitatory synaptic input recruited onto the motor neuron by the noxious stimulus. We found that both of these explanations contribute to the firing pattern of the ink gland motor neurons.

The L14 motor neurons are unusual in exhibiting antiadaptive firing behavior. A constant-amplitude excitatory stimulus, such as the injection of a step of depolarizing current, typically elicits spikes only after an initial pause of $1-2$ sec (Fig. 2A1). This intrinsic low-pass feature of the motor neurons results from a combination of two factors. First, voltage-clamp analysis revealed that the ink gland motor neurons have a high density of fast K^+ channels that, when activated, generate an outward current, I_A (see Connor, this volume). It takes $1-2$ sec for I_A to inactivate fully in these cells (Byrne et al. 1979; Byrne 1980a,b). Second, these neurons are unique among *Aplysia* neurons in having an exceptionally high V_R of -75 mV. This V_R is about 30 mV more hyperpolarized than that of most *Aplysia* nerve cells and ensures that the steady-state inactivation of the fast K⁺ channels will be negligible. Thus, when excitatory synaptic input is recruited onto an L14 cell, a large outward I_A is elicited that opposes the inward synaptic current for $1-2$ sec. The result is an intrinsic antiadaptive characteristic of the spike-generating mechanism in these cells (Byrne et al. 1979).

The temporal pattern of the excitatory synaptic input recruited by a noxious stimulus also contributes to the delay in the ink-release response. For a 5-sec stimulus, the initial part of the synaptic input is dominated by a conventional compound EPSP produced by an increased conductance. Later in the stimulus, the synaptic input is dominated by a decreased-conductance EPSP (Shapiro et al. 1979). This late EPSP, which is probably caused by a decrease in g_K, is generated in part by the identified cell L31 (Byrne 1980c). At least one reason for the delay in recruitment of the decreased-conductance PSP resides in the properties of the L31-L14 synapse. The EPSP generated in L14 by a burst of spikes in L31 takes a few seconds to reach its peak (Byrne 1980c).

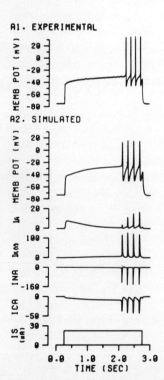

Figure 2
Comparison of experimental (A1) and simulated (A2) responses to a long-duration current pulse. (A1) When a 2.5-sec constant-current step depolarization is delivered to an ink gland motor neuron, there is an initial rapid depolarization, followed by a slow, several-second rise which terminates in a burst of four action potentials. (A2) As with the experimental data, the simulated response shows a several-second silent period before a burst of spikes is elicited. The activation and inactivation sequence of I_A appears to underlie the cell's selective response to long-duration intracellular current pulses. (Reprinted, with permission, from Byrne 1980b.)

QUANTITATIVE CONTRIBUTION OF MOTOR NEURON IONIC CONDUCTANCE MECHANISMS TO INKING BEHAVIOR

• The electrophysiological data cited above suggest that the relative insensitivity of inking behavior to brief stimuli arises from two factors: the activation of I_A, which opposes the synaptic current, and the delayed recruitment of a decreased-conductance EPSP. To test this hypothesis quantitatively, a mathematical model of the membrane of the ink gland motor neurons was constructed. This equivalent of the Hodgkin-Huxley (1952) circuit-type model includes the passive electrical properties (C_m, g_L, and E_L), the voltage-sensitive ionic channels involved in spike generation (g_{Na}, g_K, g_A, g_{Ca}, and their associated ionic batteries), plus an element that represents the lumped, time-varying properties of the synaptic channels that are activated by a sustained stimulus train to the connectives. The steady-state and kinetic properties of these membrane characteristics were determined in voltage-clamp experiments (Byrne 1980a,b). This model provides additional support for our explanation of why ink release is so insensitive to brief stimuli. Figure 2A2 shows that when the model is excited by a simulated pulse of depolarizing current, there is a strong activation of I_A that coincides with the initial delay in firing. Turning off the g_A element in the model eliminates the delay (Byrne 1980b), just as preventing I_A activation by producing steady-state inactivation or by blocking the channels with 4-AP eliminates this pause in the real cell (Byrne et al. 1979).

Figure 3 presents a comparison of the experimental results and a computer simulation of the response to synaptic input elicited by connective stimulation. In Figure 3A (the experimental case), the initial synaptic input is large, but it is ineffective in firing the cell. There is a several-second silent period or pause before an accelerating burst of action potentials is elicited. The results of the simulation are illustrated in Figure 3B. There is good agreement between the experimental data and the simulated response.

With the model, it is possible to specify the degree to which each current contributes to the firing pattern of the ink gland motor cells during the type of input that normally drives them. It is evident that I_A makes a large contribution to the cell's selective response to long-lasting stimuli. The initial synaptic current (IS) is quite large. The resultant depolarization rapidly activates I_A, which in turn shunts the synaptic current and initially prevents the cell from reaching spike threshold. With time, I_A inactivates and the synaptic current is more effective in depolarizing the cell. Eventually, V_m reaches threshold and an accelerating train of action potentials is produced. Thus, I_A can be considered to act as a braking mechanism, which ensures that only strong and long-lasting stimuli initiate firing in the ink gland motor neuron.

Although I_A appears to make a large contribution to the cell's insensitivity to brief stimuli, the build-up of synaptic current appears to play a critical role in mediating the late accelerating burst of action potentials. Figure 4 shows that by freezing the value of the synaptic current that excites the model at the value recorded at the 2.5-sec point in the stimulus, the late burst of spikes seen experimentally is eliminated. But when the delayed decreased-conductance EPSP is allowed to develop in the simulation as it does in the real cell, the late burst of spikes is generated (Fig. 3).

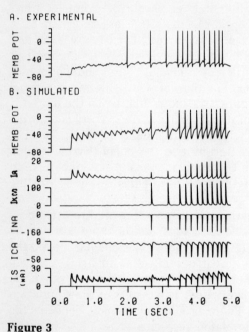

Figure 3
Comparison of experimental (A) and simulated (B) responses to synaptic input. (A) A 4.5-sec-long train of stimuli is delivered to the pleuroabdominal connectives. The initial synaptic input produced by this stimulation is ineffective in firing the cell, and there is a several-second pause before the cell begins to fire in an accelerating burst of action potentials. (B) The model was driven with the experimentally determined values of synaptic input. As with the experimental response, the initial synaptic input is ineffective in firing the cell. (Reprinted, with permission, from Byrne 1980b.)

SIMULATION OF THE NEURAL ACTIVITY UNDERLYING SHORT-TERM PLASTICITY OF INKING BEHAVIOR

● The neural correlates of the short-term behavioral modification of ink release were examined by Carew and Kandel (1977b). Subthreshold stimuli to the siphon or to the connectives that link the abdominal ganglion to the head failed to fire action potentials in the ink gland motor neurons. But when the two stimuli were delivered sequentially, a burst of spikes was produced by the second stimulus and ink was released. Thus, one neural pathway is capable of augmenting or facilitating the ability of another to fire the ink gland motor neurons and lead to the release of ink. The mechanism underlying this facilitation was examined by using a modified version of the neuronal model described above.

The simulated response when the equations were driven by the unpaired siphon input is shown in Figure 5A. There is a surge of synaptic current that produces a large initial depolarization, but no

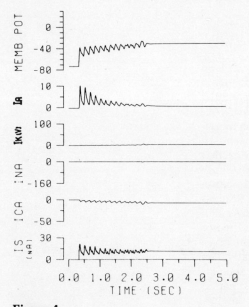

Figure 4
Role of late synaptic input in mediating firing pattern of ink gland motor neurons. Simulation of Fig. 3 was repeated under identical conditions, except that synaptic resistance and equilibrium potential were held fixed after 2.5 sec. With the synaptic input held fixed, the accelerating burst is eliminated. (Reprinted, with permission, from Byrne 1980b.)

A SIPHON ALONE

B CONNECTIVES + SIPHON

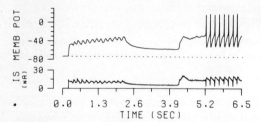

Figure 5
Simulation of the firing pattern of ink gland
motor neurons. (A) Simulation of siphon
nerve input. Synaptic equilibrium potential
and synaptic conductance were analyzed
when ink gland motor neurons were driven
by input from the siphon nerve. These estimates were then used to drive the model that
describes the nonsynaptic membrane (Byrne
1980b). The results of a computer simulation
of the equations are illustrated. (B) Simulation of connective input paired with siphon
input. Synaptic current to ink gland motor
neurons was analyzed while connectives
and siphon were stimulated electrically. The
2-sec-duration stimulus trains were separated by 2 sec. The estimates of the synaptic
conductance and equilibrium potential during the pairing sequence were used to drive
the equations. The simulation illustrates
that the connective input is also subthreshold. However, when the siphon input is delivered, eight spikes are produced as compared with no spikes when the siphon input
is unpaired (A). (Reprinted, with permission, from Byrne 1980d.)

spikes are initiated in the cell. Figure 5B illustrates the results of the simulation when connective input is paired with siphon input. The connective input depolarizes the cell rapidly, but, as with the unpaired siphon input (Fig. 5A), no spikes are produced. When the siphon input follows the connective input 2 sec later, however, a burst of spikes is produced. Thus, the simulations are consistent with the experimental observations of Carew and Kandel (1977b) that activity in one stimulus pathway augments the ability of another pathway to fire the ink gland motor neurons. With the simulation, it is possible to pinpoint the mechanism underlying this modification. The connective input depolarizes the cell not only for the 2-sec period of the train, but also for a significant period after the train (Fig. 5B; see also Carew and Kandel 1977b). This depolarization, in turn, is due to a sustained increase in synaptic current (Fig. 5B, lower trace). As a result, when the siphon pathway is activated, the synaptic current triggered by the siphon stimulus summates with the residual synaptic current elicited by head stimulation, and the previously subthreshold synaptic current then is capable of driving the cell to threshold. Thus, the sustained synaptic current appears to be linked to the features of the behavioral modification. The sustained increase in synaptic current is mediated, at least in part, by the identified neuron L31 (see above), which produces a slowly activating and long-lasting decreased-conductance EPSP in the ink gland motor neurons (Byrne 1980c).

As pointed out by Carew and Kandel (1977b), this augmentation of the effect of the second stimulus is not due simply to the depolarization and cannot be accounted for entirely by the residual depolarization, since artificially depolarizing the cell is ineffective in producing the enhancement effect. The simulations also confirm this finding. The simulation of Figure 5A was repeated, but 4 sec prior to and during the synaptic input the model was driven by a constant current sufficient to depolarize the motor neuron to −57 mV (the final value achieved just prior to the arrival of the siphon input in Fig. 5B). The siphon input, as in Figure 5A, now produced only three spikes. Thus, although there is some summation of depolarizations in this example, it is not as effective in augmenting the second response as when the depolarization is produced by prior connective stimulation. Presumably, the synaptically evolved depolarization is more effective because it is produced in part by a decreased-conductance mechanism. This is a particularly effective way of exciting a cell because it reduces the effective g_L. This decrease in input conductance in turn causes other synaptic inputs to be more effective. The connective stimulus may also produce a presynaptic facilitation of the siphon input (see Klein, this volume).

SUMMARY

• The neural circuit that controls ink release in *Aplysia* has been worked out in some detail (Byrne 1980c). By just looking at this circuit, however, it is not possible to predict the temporal aspects of inking behavior described above. It is only when one takes into account the dynamic properties of individual nerve cells that explanations emerge for the relations between stimulus duration or pattern and the inking response. These explanations assign important functions to two of the recently discovered types of conductance channels found in nerve cells: The fast K^+ channels, which in this case exhibit very little steady-state inactivation, are found to endow the inking system with "inertia," ensuring that ink release will occur only in response to relatively long-lasting stimuli. Thus, in addition to their more common role in controlling repetitive firing properties (see Connor, this volume), the fast K^+ channels can also act, when present in cells with high V_Rs, to produce a delay. The relatively slow onset of the decreased-conductance synaptic potentials elicited in the L14 cells by a noxious stimulus also contributes to the selective responsiveness of these cells to long-lasting stimuli. In addition, the slow decay of this decreased-conductance EPSP partially accounts for the short-lasting behavioral plasticity exhibited by the inking system.

In this presentation, as well as in that by Klein (this volume), the biophysical properties of identified neurons in well-defined neural networks that control discrete behaviors have been examined. It has been possible to relate specific biophysical properties of a neuron to its functional role in the generation of behavior. As this approach is extended to other behavioral systems, it is likely that different behaviorally relevant functions will emerge for the other recently described species of ionic channels.

REFERENCES

Byrne, J.H. 1980a. Analysis of ionic conductance mechanisms in motor cells mediating inking behavior in *Aplysia californica. J. Neurophysiol.* **43**: 630.

————. 1980b. Quantitative aspects of ionic conductance mechanisms contributing to firing pattern of motor cells mediating inking behavior in *Aplysia californica. J. Neurophysiol.* **43**: 651.

————. 1980c. Neural circuit for inking behavior in *Aplysia californica. J. Neurophysiol.* **43**: 896.

————. 1980d. Simulation of the neural activity underlying a short-term modification of inking behavior in *Aplysia. Brain Res.* (in press).

Byrne, J., E. Shapiro, N. Dieringer, and J. Koester. 1979. Biophysical mechanisms contributing to inking behavior in *Aplysia. J. Neurophysiol.* **42**: 1233.

Carew, T.J. and E.R. Kandel. 1977a. Inking in *Aplysia californica*. I. Neural circuit of an all-or-none behavioral response. *J. Neurophysiol.* **40:** 692.

———. 1977b. Inking in *Aplysia californica*. III. Two different synaptic conductance mechanisms for triggering the central program for inking. *J. Neurophysiol.* **40:** 721.

Hodgkin, A.L. and A.F. Huxley. 1952. A quantitative description of membrane current and its application to conduction and excitation in nerve. *J. Physiol. (Lond.)* **117:** 500.

Shapiro, E., J. Koester, and J.H. Byrne. 1979. *Aplysia* ink release: Central locus for selective sensitivity to long duration stimuli. *J. Neurophysiol.* **42:** 1223.

The Neuronal Pacemaker Cycle

ANTHONY L.F. GORMAN,
ANTON HERMANN,*
and MARTIN V. THOMAS†

Department of Physiology
Boston University School of Medicine
Boston, Massachusetts 02118

• One of the characteristic features of pacemaker cells is their ability to generate rhythmic activity. Such cells are found in heart and smooth muscle, in invertebrate nervous systems, and in cultured mammalian neurons (see Carpenter 1978; Berridge et al. 1979; Gähwiler and Dreifuss 1979). The origin of the endogenous oscillatory changes in V_m that are recorded from these cells is not fully understood. Most of the mechanisms that have been proposed involve either cyclic changes in time- and voltage-dependent membrane conductance systems or in the cyclic modulation of active processes within or coupled to the cell membrane. Summaries of these ideas are provided in two recent symposia (Carpenter 1978; Berridge et al. 1979). The pacemaker cells found in certain molluscan neuronal ganglia have been studied extensively, and there is now a large body of information about pacemaker activity in several of these neurons (Eckert and Lux 1976; Gola 1976; Meech 1976; Gorman and Thomas 1978; Gulrajani and Roberge 1978; Carnavale and Wachtel 1980; T. Smith, this volume) that can be used to formulate a reasonable hypothesis to explain the ionic basis of the pacemaker cycle. The data suggest that there is a slow inward current during the depolarizing phase of the pacemaker cycle. This inward current is the result of a voltage-dependent change in g_m in a channel that is insensitive to TTX (Gola 1974; Wilson and Wachtel 1974; Smith et al. 1975; Eckert

*Present address: University of Konstanz, Department of Biology, D-7750 Konstanz, Federal Republic of Germany.
†Present address: Shell Biosciences Laboratory, Sittingbourne, Kent ME9 8AG, England.
This work was supported by NIH grant NS11429.

and Lux 1976; Johnston 1976). Following the slow inward current and the resultant spike burst it produces, there is a subsequent outward current, which causes the hyperpolarizing phase of the cycle (Junge and Stephens 1973; Gola 1974; Meech 1974; Gorman and Thomas 1978). The current across the molluscan soma membrane can be separated into a variety of components (Stevens, this volume), but it is not clear which of these is crucial for pacemaker activity. Previously, we showed (Thomas and Gorman 1977; Gorman and Thomas 1978) in the bursting pacemaker neuron R15 of *Aplysia* that an increase in intracellular Ca^{++} occurred during the depolarizing phase of the cycle and could cause, or assist in causing, the subsequent hyperpolarizing phase by activating an outward I_K. Evidence is given below that suggests that the pacemaker cycle of cell R15 in the abdominal ganglion of *Aplysia californica* depends on two opposing membrane conductance systems, one of which controls the inward movement of Ca^{++} and the other the outward movement of K^+, which is linked by changes in $[Ca^{++}]_i$.

INWARD MOVEMENT OF Ca^{++} DURING THE PACEMAKER CYCLE

• The relationship between changes in $[Ca^{++}]_i$, as measured by the absorbance change of the metallochromic indicator dye arsenazo III (Gorman and Thomas 1978), and changes in the V_m of cell R15 during pacemaker activity is shown in Figure 1, A and B. The V_m of cell R15 undergoes a series of cyclic variations consisting of a slow depolarization lasting some $10-15$ sec, which, when it reaches a threshold of about -40 mV to -35 mV, gives rise to a reasonably constant number of action potentials. The depolarizing phase then gives way to a longer ($15-25$ sec) hyperpolarizing phase that reaches a maximum a few seconds after each burst of action potentials and subsequently declines. There is an increase in dye absorbance during the depolarizing phase of the cycle (Fig. 1A), indicating an increase in $[Ca^{++}]_i$, and a subsequent decrease during the hyperpolarizing phase of the pacemaker cycle. The absorbance change in different cells corresponds to an increase in $[Ca^{++}]_i$ of about $30-100$ nM. The increase begins at the start of each burst and is augmented by each action potential during the burst (Fig. 1B). The increase in $[Ca^{++}]_i$ precedes and overlaps the postburst hyperpolarization. Any increase in membrane depolarization during the depolarizing phase of the cycle increases the magnitude and duration of the change in $[Ca^{++}]_i$ and of the postburst hyperpolarization (Fig. 1A). Evidence obtained by the use of metabolic inhibitors suggests that this increase in $[Ca^{++}]_i$ occurs as a result of Ca^{++} influx through the membrane rather than from release of intracellular stores (Gorman and Thomas 1978).

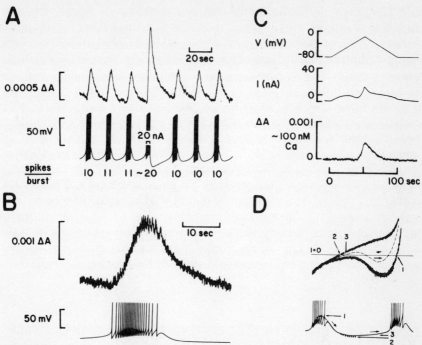

Figure 1

(A) Changes in [Ca++]i (*upper trace*) during pacemaker activity (*lower trace*) in the *Aplysia* R15 neuron. The effect of artificially augmenting a burst of action potentials on [Ca++]i measured with arsenazo III. A 20-nA depolarizing current was passed for the period indicated, approximately doubling the number of action potentials and increasing the amplitude and duration of the subsequent hyperpolarization. (Reprinted, with permission, from Gorman and Thomas 1978.) (B) As above, but from a different cell in which the change in [Ca++]i with each action potential can be seen more clearly (A.L.F. Gorman and M.V. Thomas, unpubl.). (C) Changes in [Ca++]i (*lower trace*) and I_m (*middle trace*) produced by triangular ramp depolarizations (*upper trace*) from a V_h of −80 mV in cell R15 (A.L.F. Gorman and M.V. Thomas, unpubl.). (D) I_m generated by a depolarizing triangular ramp change in V_m of cell R15 (*above*). The currents during the depolarizing ramp have been superimposed on those recorded during the repolarizing ramp. Typical pacemaker activity recorded from cell R15 (*below*). The points identified by numbers are correlated with different phases of the underlying dynamic I-V relation. (Modified from Gola 1974.)

The changes in I_m underlying the pacemaker cycle can be examined experimentally by slowly depolarizing and repolarizing the cell membrane under voltage-clamp conditions using a triangular voltage command (Gola 1974). As shown in Figure 1C, an inward current develops at about −50 mV when V_m is depolarized from a V_h

of −80 mV, but the current becomes outward at V_m values more positive than about −25 mV. This inward current does not reappear however, when the membrane is repolarized. The inward-outward current pattern during depolarization imparts a region of negative-slope conductance to the dynamic I-V relation of cell R15, and the partial or complete suppression of the inward current during repolarization causes the relation to be asymmetrical (Fig. 1D). An I-V relation that is characterized by a region of negative-slope conductance and hysteresis is inherently unstable and has been correlated by several groups (Gola 1974; Wilson and Wachtel 1974; Smith et al. 1975; Eckert and Lux 1976; Johnston 1976; T. Smith, this volume) with oscillatory membrane activity in pacemaker cells. Gola (1974, 1976) has presented a qualitative model that summarizes these findings and accounts for the endogenous oscillation of V_m in terms of two conductance systems that change with voltage and with time (Fig. 1D). The identity of the ionic channels that control these two conductances and their interrelationship during pacemaker activity is given below.

IONIC BASIS OF THE SLOW INWARD CURRENT

• As shown in Figure 1C, $[Ca^{++}]_i$ increases when the membrane is depolarized beyond −45 mV to −40 mV and the increase declines slowly during membrane repolarization. This change in $[Ca^{++}]_i$ is abolished by procedures that block Ca^{++} influx, for example, removal of external Ca^{++} or the presence of external divalent and trivalent blocking cations. These data confirm that Ca^{++} enters through voltage-dependent channels that are open within the V_m range corresponding to the depolarizing phase of the pacemaker cycle (Eckert and Lux 1976). In addition, Figure 1A shows that the change in $[Ca^{++}]_i$ is correlated well with the appearance of an outward current and is consistent with the activation of a component of I_K by an increase in the $[Ca^{++}]_i$ during the pacemaker cycle (Thomas and Gorman 1977; Gorman and Thomas 1978; Meech, this volume).

The ionic basis of the slow inward current in cell R15 is a matter of controversy. In Helix pacemaker neurons, there is evidence that the slow inward current is carried predominantly by Ca^{++} (Eckert and Lux 1976), but in other molluscan pacemaker neurons, including cell R15, it has been suggested that the inward current is carried by Na^+ through a channel that is not affected by TTX (Smith et al. 1975, T. Smith, this volume), or by both Na^+ and Ca^{++} (Gola 1976; Johnston 1976). It has been shown that in Helix neurons K^+ efflux occurs simultaneously with the appearance of the slow inward current (Heyer and Lux 1976), which suggests that any analysis of the

ionic basis of the slow inward current is likely to be in serious error
unless outward I_K can be suppressed fully. Techniques for suppressing the outward I_K in molluscan neurons are available (Connor 1979;
Ehile and Gola 1979; Hermann and Gorman 1979). The inward current produced by a slow depolarizing ramp in an R15 cell that had
been injected with sufficient EGTA to prevent a change in $[Ca^{++}]_i$
following Ca^{++} influx, thereby suppressing $I_{K(Ca)}$, is shown in Figure
2A. The cell was maintained in an external solution containing 100
mM TEA and 5 mM 4-AP to suppress $I_{K(V)}$ and I_A and in sufficient TTX
(5×10^{-5} M) to block the fast inward I_{Na}. The remaining current is
composed of an inward current and the leakage current. The inward
current shown in Figure 2A is substantially larger than the current
recorded from the uninjected cell in normal artificial seawater (Fig.
1C) and suggests that, under normal conditions, the inward current is

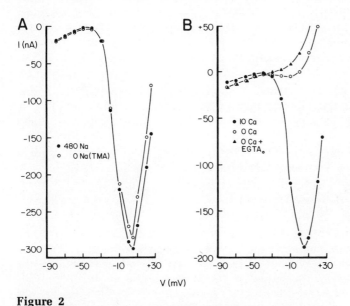

Figure 2
I-V relation produced by a depolarizing ramp change in
V_m. (A) The effects of the removal of external Na$^+$ on the
I-V relation. The cell was injected previously with EGTA
iontophoretically for 5 min at an intensity of 500 nA and
maintained in an external solution containing 100 mM
TEA, 5 mM 4-AP, and 5×10^{-5} M TTX. Na$^+$ was replaced
completely by TMA on an equimolar basis. (B) Effects of
the removal of external Ca^{++} on the I-V relation of a
different R15 cell. As above, but external Ca^{++} was replaced by Mg^{++} on a 1:1 basis. In addition, 2 mM external
EGTA was used to chelate any residual Ca^{++} (A.L.F.
Gorman and A. Hermann, unpubl.).

masked almost completely by overlapping I_K. Complete replacement of external Na^+ with tetramethylammonium (TMA), which has no effect on $I_{K(Ca)}$ and $I_{K(V)}$ at external concentrations up to 500 mM (Hermann and Gorman 1979), had almost no effect on the inward current, whereas the current was abolished by the complete replacement of external Ca^{++} by Mg^{++} (Fig. 2 A, B). The results indicate that the slow inward current in cell R15 is carried almost exclusively by Ca^{++} and not by Na^+.

K^+ FLUX DURING THE PACEMAKER CYCLE

• The relationship between changes in $[K^+]_e$ and changes in the V_m of cell R15 during the pacemaker cycle is shown in Figure 3A. There is an increase in $[K^+]_e$ during the depolarizing phase of the cycle and a subsequent decline during the hyperpolarizing phase. The increase begins at the start of each burst and reaches a maximum during the depolarizing phase of the cycle. The decline is particularly rapid at the start of the hyperpolarizing phase (see also Lux and Heyer 1975), but this is expected because the efflux of K^+ through the membrane ceases when V_m moves toward E_K during the postburst hyperpolarization. Figure 3A also shows that an increase in membrane depolarization during the pacemaker cycle increases the magnitude and duration of the change in $[K^+]_e$ and of the postburst hyperpolarization. This result is consistent with the increase in $[Ca^{++}]_i$ under similar conditions (Fig. 1A) and with the ·increase in $I_{K(Ca)}$ following an increase in $[Ca^{++}]_i$ (Fig. 3C). The change in $I_{K(Ca)}$ and in $[K^+]_e$ occurs during the small inward current that develops during the depolarizing phase of the triangular ramp command and overlaps the subsequent outward current. The magnitude of the change in $[K^+]_e$ during this small voltage change suggests that a substantial fraction of the K^+ efflux during the depolarizing phase of the pacemaker cycle is not associated with the discharge of action potentials.

PROPERTIES OF $I_{K(Ca)}$ IN CELL R15

• $I_{K(Ca)}$ is studied most easily by direct injections of Ca^{++} (Gorman and Hermann 1979; Gorman and Thomas 1980; Meech, this volume). There is strong evidence that the current activated by iontophoretic or pressure injection of Ca^{++} represents a selective increase in g_K (Brown and Brown 1973; Gorman and Hermann 1979; Meech, this volume). The magnitude of $I_{K(Ca)}$ is determined by the intensity and duration of the iontophoretic injection current and the position of the injection electrode with respect to the inner membrane surface, as well as by V_h (Gorman and Hermann 1979). A normalized plot of the change in $[Ca^{++}]_i$, as measured by arsenazo III, and $I_{K(Ca)}$, produced by Ca^{++} injection, vs time is shown in Figure 3C. The rise and decline of $I_{K(Ca)}$ parallels the average change in $[Ca^{++}]_i$ but is slower because of the time required for Ca^{++} to arrive at the inner membrane surface (Gorman and Thomas 1980; S. Smith, this volume). The relation between the peak $I_{K(Ca)}$ and the maximum change in absorbance produced during Ca^{++} injection at various intensities, but constant

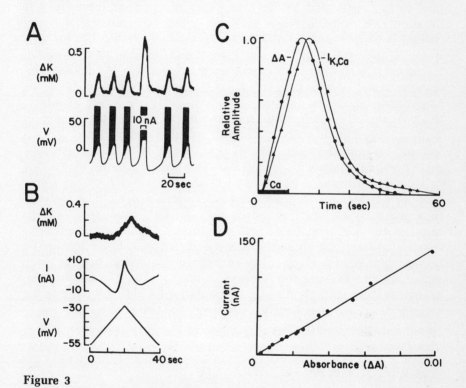

Figure 3
(A) $[K^+]_e$ change (*upper trace*) measured with a K^+-sensitive external micro-electrode during pacemaker activity (*lower trace*) in cell R15. The effects of artificially augmenting the depolarizing phase of the pacemaker cycle on $[K^+]_e$ and on the subsequent hyperpolarization is shown. The augmentation was produced by a 10-nA depolarization for the time indicated. (B) $[K^+]_e$ change (*upper trace*) and I_m (*middle trace*) produced by a triangular ramp depolarization. (Reprinted, with permission, from Gorman et al. 1981.) (C) The normalized absorbance change ($\Delta A/\Delta A_{max}$) and the normalized value of $I_{K(Ca)}$ ($I_{K(Ca)}/I_{K(Ca)max}$) produced by a 10-sec, 200-nA Ca^{++} injection is plotted vs time. (D) Plot of the outward I_K response vs the change in absorbance produced by constant-duration (30 sec) Ca^{++} injection at different intensities. An absorbance change of 0.01 units corresponds to an increase in internal Ca^{++} of about 1.4 μM. V_h was −40 mV in A and B. (Reprinted, with permission, from Gorman and Thomas 1980.)

duration, is shown in Figure 3D. The excellent linearity of the relation is apparent and suggests that one Ca^{++} ion binds to each internal site that activates a $g_{K(Ca)}$ channel. The ability of a divalent cation to activate g_K is related to its radius. The order of affinity for activation of I_K is $Ca^{++} > Cd^{++} > Hg^{++} > Sr^{++} > Mn^{++} > Fe^{++}$, and it suggests that the site can accommodate ionic radii between about 0.76 Å and 1.13 Å (Gorman and Hermann 1979).

The relation between $I_{K(Ca)}$ and V_m is nonlinear. In Figure 4A a plot of $I_{K(Ca)}$ activated by Ca^{++} injection of constant intensity and duration over a wide range of V_m is provided, showing that the current increases substantially at more positive V_m levels. E_{rev} for $I_{K(Ca)}$ was −68 mV in this cell. I_K was inward and small at more negative V_m levels and outward and large at more positive V_m levels. The instantaneous I-V relation for $I_{K(Ca)}$ is reasonably linear, so $g_{K(Ca)}$ can be estimated from the relation $g_{K(Ca)} = I_{K(Ca)} / (V_h - E_K)$ (see also Lux, this volume). The conductance increases e-fold for a 24-mV change in V_m (Fig. 4B). This is smaller than the voltage dependence of $I_{K(V)}$ (e-fold for 12.5 mV between −20 mV and 0 mV), which is activated at more positive levels of V_m in cell R15. One mechanism by which this voltage dependence might occur is that the affinity of the Ca^{++}-binding site for Ca^{++} is greater at positive levels of V_m. The physical basis for such an effect would be that the Ca^{++}-binding site is (in terms of the electrical field) part way through the membrane rather than at the inner membrane surface (Gorman and Thomas 1980). The voltage dependence of $g_{K(Ca)}$ has an important consequence for pacemaker activity because the increase in $[Ca^{++}]_i$ (Fig. 1) becomes progressively more important as V_m is depolarized and thereby limits the magnitude of the depolarizing phase of the cycle.

A MODEL OF THE PACEMAKER CYCLE

• The data given above and elsewhere (Gorman et al. 1981) is summarized in Figure 5, which provides a qualitative picture of those changes that are necessary for maintaining pacemaker activity in cell R15. The pacemaker cycle is composed of three separate but linked systems: (1) a voltage-dependent Ca^{++} current, I_{Ca}, (2) $[Ca^{++}]_i$, and (3) $I_{K(Ca)}$. Depolarization causes the activation of an inward current carried primarily by Ca^{++} through voltage-dependent Ca^{++} channels, which are slow to inactivate (Eckert and Lux 1976; Kostyuk and Krishtal 1977; Connor 1979; Ehile and Gola 1979; Brown; Tillotson; both this volume). This current is regenerative and provides the additional depolarization necessary to bring V_m to the threshold for action potential initiation and to maintain it above this level. Ca^{++} entry and accumulation occur throughout the slow depolarization

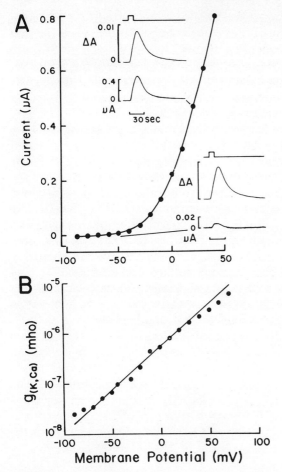

Figure 4

Effects of V_m on $I_{K(Ca)}$ and $g_{K(Ca)}$. (*A*) The differential absorbance change and $I_{K(Ca)}$ (*inset*) produced by two intracellular Ca^{++} injections (10 sec, 100 nA) at a V_h of -50 mV and $+20$ mV, respectively, in Ca^{++}-free artificial seawater. (*B*) Logarithmic plot of $g_{K(Ca)}$ activated by a constant intracellular Ca^{++} injection vs membrane V_h in normal artificial seawater. $g_{K(Ca)}$ was activated by a constant 10-sec, 100-nA electrophoretic injection of Ca^{++} into the cell at each V_m (Gorman and Thomas 1980).

and during each action potential. The subsequent rise in [Ca^{++}]$_i$ activates K$^+$ channels, which carry an opposing outward current. This activation depends both on the amount of free intracellular Ca^{++} that accumulates during the depolarizing phase of the cycle and on V_m.

When the outward current exceeds the inward current, V_m becomes more negative, thereby beginning the hyperpolarizing phase of the cycle. This hyperpolarization has several consequences: The outward flux of K^+ diminishes because of the reduced driving force acting on K^+ movement; the voltage-dependent Ca^{++} channels are turned off; and the Ca^{++} that accumulated is slowly removed by internal sequestration and membrane extrusion systems causing the Ca^{++}-activated K^+ channels to close; the closure of K^+ channels, in turn, results in a slow depolarization that restarts the cycle.

There are several other currents that occur in the molluscan neuronal membrane and contribute to pacemaker activity, but their role in maintaining the pacemaker cycle appears to be secondary rather than direct. These currents do play a direct role in determining the shape of the action potential and in the control of the frequency of action potential discharge during the burst, thereby contributing indirectly to the change in $[Ca^{++}]_i$. The fast inward current, which is carried by Na^+ and can be blocked by external TTX (Stevens, this volume), contributes to the depolarizing phase of each action potential. Similarly, $I_{K(V)}$ contributes to the repolarization of the action potential and, therefore, indirectly to the length of time Ca^{++} chan-

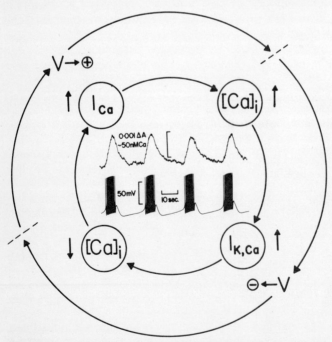

Figure 5
Diagrammatic representation of the pacemaker cycle in cell R15. See text for further details. (Reprinted, with permission, from Gorman et al. 1981.)

nels remain open (see Thompson, this volume). Finally, the fast outward I_K (I_A) helps to control the frequency of action potentials during the burst (Connor, this volume) and, therefore, the amount of Ca++ that enters the cell.

Our picture of the events underlying pacemaker activity (Fig. 5) is more complex than previous explanations, which assume that membrane oscillations are controlled by cyclic changes in time- and voltage-dependent conductances or by metabolic processes. One of the advantages of viewing pacemaker oscillations as being determined by a cycle of events is that a system of this type can be used to account for the effects of the numerous and diverse agents that have been used to alter or abolish pacemaker activity. The cycle can be interfered with at any point; for example, it is possible to disrupt or alter the cycle by reducing or increasing Ca++ influx through Ca++ channels by changing the intracellular regulation of Ca++ or by reducing or increasing K+ efflux through Ca++-activated K+ channels. This can be done not only directly, but also indirectly by altering those currents that control the amplitude, the time course, and the frequency of action potential discharge. Moreover, it is possible to induce, as well as to modify or abolish, pacemaker activity in cells that contain voltage-dependent Ca++ and Ca++-activated K+ channels by using agents that work primarily upon membrane or upon intracellular Ca++ regulatory sites.

REFERENCES

Berridge, M.J., P.E. Rapp, and J.E. Treherne, eds. 1979. "Cellular oscillators." *J. Exp. Biol.* **81.**

Brown, A.M. and H.M. Brown. 1973. Light response of a giant *Aplysia* neuron. *J. Gen. Physiol.* **62:** 239.

Carnevale, N.T. and H. Wachtel. 1980. Two reciprocating current components underlying slow oscillations in *Aplysia* bursting neurons. *Br. Res. Rev.* **2:** 45.

Carpenter, D.O. 1978. Cellular pacemakers. *Fed. Proc.* **37:** 2125.

Connor, J.A. 1979. Calcium current in molluscan neurons: Measurement under conditions which maximize its visibility. *J. Physiol. (Lond.)* **286:** 41.

Eckert, R. and H.D. Lux. 1976. A voltage-sensitive persistent calcium conductance in neuronal somata of *Helix. J. Physiol. (Lond.)* **254:** 129.

Ehile, E. and M. Gola. 1979. A slowly inactivating calcium current in molluscan neurons. I. Slow currents during long-lasting voltage clamp pulses. *Comp. Biochem. Physiol.* **64A:** 213.

Gähwiler, B.H. and J.J. Dreifuss. 1979. Physically firing neurons in long-term cultures of the rat hypothalamic supraoptic area: Pacemaker and follower cells. *Brain Res.* **177:** 95.

Gola, M. 1974. Neurones a ondes-salves des mollusques. Variation cycliques lentes des conductances ioniques. *Pfluegers Archiv. gesamte Physiol. Menschen Tiere* **352:** 17.

————. 1976. Electrical properties of bursting pacemaker neurones. In *Neurobiology of invertebrates, gastropoda brain* (ed. J. Salanki), p. 381. Akademiai Kiado, Budapest.

Gorman, A.L.F. and A. Hermann. 1979. Internal effects of divalent cations on potassium permeability in molluscan neurones. *J. Physiol. (Lond.)* **296:** 393.

Gorman, A.L.F. and M.V. Thomas. 1978. Changes in the intracellular concentration of free calcium ions in a pacemaker neurone, measured with the metallochromic indicator dye arsenazo III. *J. Physiol. (Lond.)* **275:** 357.

————. 1980. Potassium conductance and internal calcium accumulation in a molluscan neuron. *J. Physiol. (Lond.)* (in press).

Gorman, A.L.F., A. Hermann, and M.V. Thomas. 1981. Intracellular calcium and the control of neuronal pacemaker activity. *Fed. Proc.* (in press).

Gulrajani, R.M. and F.A. Roberge. 1978. Possible mechanisms underlying bursting pacemaker discharges in invertebrate neurons. *Fed. Proc.* **37:** 2146.

Hermann, A. and A.L.F. Gorman. 1979. External and internal effects of tetraethylammonium on voltage-dependent and Ca-dependent K⁺ current components in molluscan pacemaker neurons. *Neurosci. Lett.* **21:** 87.

Heyer, C.B. and H.D. Lux. 1976. Properties of a facilitating calcium current in pacemaker neurones of the snail, *Helix pomatia*. *J. Physiol. (Lond.)* **262:** 319.

Johnston, D. 1976. Voltage clamp reveals basis for calcium regulation of bursting pacemaker potentials in *Aplysia* neurons. *Brain Res.* **107:** 418.

Junge, D. and C.L. Stephens. 1973. Cyclic variation of potassium conductance in a burst-generating neurone in *Aplysia*. *J. Physiol. (Lond.)* **235:** 155.

Kostyuk, P.G. and O.A. Krishtal. 1977. Separation of sodium and calcium currents in the somatic membrane of mollusc neurones. *J. Physiol. (Lond.)* **270:** 545.

Lux, H.D. and C.B. Heyer. 1975. Fast K⁺ activity determinations during outward currents of the neuronal membrane of *Helix pomatia*. *Bioelectrochem. Bioenerg.* **3:** 169.

Meech, R.W. 1974. The sensitivity of *Helix aspersa* neurones to injected calcium ions. *J. Physiol. (Lond.)* **273:** 259.

————. 1976. Intracellular calcium and the control of membrane permeability. *Symp. Soc. Exp. Biol.* **30:** 161.

Smith, T.G., J.L. Barker, and H. Gainer. 1975. Requirements for bursting pacemaker potential activity in molluscan neurones. *Nature* **253:** 450.

Thomas, M.V. and A.L.F. Gorman. 1977. Internal calcium changes in a bursting pacemaker neuron measured with arsenazo III. *Science* **196:** 531.

Wilson, W.A. and H. Wachtel. 1974. Negative resistance characteristic essential for the maintenance of slow oscillations in bursting neurons. *Science* **186:** 932.

Ion Permeation in Synaptic Channels

Based on a presentation by

ALAIN MARTY*

Laboratoire de Neurobiologie
Ecole Normale Supérieure
Paris, France 75230

• The properties of voltage-gated and Ca^{++}-activated ion channels have been discussed in considerable detail earlier in this volume. This and subsequent presentations will examine some of the properties of channels that are activated by neurotransmitter substances. Until recently, voltage-gated and transmitter-gated channels appeared to be separate entities. Voltage-gated channels were believed to be insensitive to transmitters. Similarly, transmitter-gated channels were thought to be insensitive to voltage.

These views, however, may be oversimplified. At the frog neuro-muscular junction, the mean open time of ACh-sensitive channels increases as V_m is made more negative (Kordaš 1969; Magleby and Stevens 1972a; Anderson and Stevens 1973). A similar effect is found for the excitatory response to ACh in *Aplysia* (Marty et al. 1976; Ascher et al. 1978a) and in the eel electroplaque (Sheridan and Lester 1975). To explain this phenomenon, Magleby and Stevens (1972b) suggested that the ACh receptor protein bears a net dipole moment and that variations of the channel closing (and opening) probabilities with voltage result from changes in the orientation of this dipole in the externally applied electric field. On the other hand, Kordaš (1969) and Adams (1976a) suggested that ACh binding is favored by cell hyperpolarization and that the channel-open time increases as a result of the higher affinity of ACh.

Another explanation for the voltage-dependence of ACh channels is suggested by the hypothesis that the binding to channel sites of

---

*Work presented here was done in collaboration with D. Marchais.
The report of this presentation was prepared by J. Strong.

permeant ions in the extracellular solution impedes channel closing (Ascher et al. 1978a). It is expected that such an effect should be favored by hyperpolarization, so that the channel-open time should increase at negative values of V_m. This mechanism has the advantage of accounting for two other experimental results (Ascher et al. 1978a), namely, that (1) the opening probability of ACh-sensitive channels does not appear to be affected significantly by V_m (Neher and Sakmann 1975; Sheridan and Lester 1975; Ascher et al. 1978a) and (2) in *Aplysia*, the mean channel-open time has a stronger voltage dependence in an external solution containing only Mg^{++} as a permeant ion than it does in normal seawater (Ascher et al. 1978a). The possibility that in *Aplysia* neurons a voltage-dependent binding of permeant ions to ACh-sensitive channels could reduce the closing rate of these channels is assessed in this presentation.

TIME CONSTANTS OF CHANNEL CLOSING DEPEND ON THE PERMEANT SPECIES

• Voltage-jump relaxation experiments were performed to assess the contribution of ion permeation to channel-closing properties. Neurons perfused in seawater were voltage-clamped to various levels, and the nonsynaptic currents were measured (e.g., the linear steplike currents in Fig. 1, top left). ACh was then applied and the identical clamp sequence was repeated (exponential traces, Fig. 1, left). The differences between the total currents measured in seawater and in the presence of ACh are the currents flowing through the channels opened by ACh. The differences in the exponential waveforms are assumed to reflect the normal voltage-dependence of the channel-closing process (see Neher and Sakman 1975; Neher and Stevens 1977). The currents seen for the same voltage steps, after Na^+ and Ca^{++} have been replaced isotonically with Mg^{++}, are shown in Figure 1 (right). The currents in Mg^{++} are smaller, and the time constants are slower and much more dependent on voltage. These results are seen more clearly in Figure 2, where results from similar experiments with several different monovalent and divalent cations are summarized. The figure shows that, at a given voltage, the value of the relaxation time constant (τ) can vary by almost half an order of magnitude as the ionic species is varied. The value of τ depends on voltage, becoming smaller as the membrane is depolarized. The voltage dependency is much stronger in the case of the divalent ions. The maximal slopes of the τ-V curves are e-fold/25 mV in the case of the divalent ions and e-fold/45−50 mV for the monovalent ions.

A SIMPLE MODEL OF ION-CHANNEL INTERACTION

• Results such as those seen in Figure 2 could be explained by assuming that the ACh channel cannot close when it contains a permeating ion. Since hyperpolarization would tend to drive cations from the external solution into the channel, there would be a greater probability at more hyperpolarized voltages that an ion was in the channel and, hence, a longer mean channel-open time would result. Since hyperpolarization would be more effective in moving divalent ions rather than monovalent ions into the channel, the stronger dependence of τ on voltage in the presence of divalent ions would also be predicted by this theory. A simple extension of the two-state model, in which the channnel can be either fully open or fully closed (see Neher and Stevens 1977), can accommodate the postulate that ions in the channel hold it in an open state. An extended version of the model, similar to the one used by Adams (1979b) or Ascher et al. (1978b) to describe blockers of the ACh channel, can be described by the scheme

$$R_c \underset{\alpha}{\overset{\beta}{\rightleftharpoons}} R_o \rightharpoonup R_o \cdot I \tag{1}$$

where R_o represents the open conformation of the receptor-channel complex, R_c the closed conformation, β the rate constant for channel opening, and α the rate constant for channel closing. $R_o \cdot I$ represents

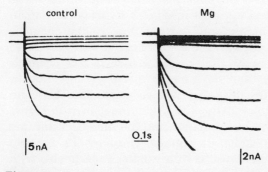

Figure 1
Voltage-clamp currents measured during relaxation experiment in Mg^{++}-seawater. Voltage jumps were applied from a V_h of -40 mV to test V_ms of -60 mV, -80 mV, -100 mV, and -120 mV in normal seawater (*left*) and in an isotonic $MgCl_2$ solution (*right*). Background currents obtained in the absence of ACh (*upper traces*) display linear behavior with voltage, whereas currents obtained in the presence of ACh (*lower traces*) increase exponentially following the voltage jumps (same iontophoretic dose in control and Mg^{++} solutions). The ACh-induced current at -120 mV went off scale (*right panel*). Temperature, 12°C.

the open channel with an ion in it, the state in which the channel can not enter the closed state. The second reaction is assumed to be rapid and at equilibrium, and therefore simple two-state kinetics are still predicted, with the modification that the observed rate constant α' will be equal to $\alpha \cdot [R_o/(R_o + R_o \cdot I)]$. The system is not described completely by the above scheme, however, since the ion, I, not only "binds" to the channel, but also, unlike a conventional blocker, permeates through the channel. Permeation through the channel can be described by the Eyring model shown in Figure 3, which shows the energy profile of one type of ion (Na^+) as it traverses the channel. The ion hops into and out of a binding site in the interior of the channel, and rate constants are determined by the height of the two barriers according to the Eyring rate theory (Laeuger 1973). From experimental observations and certain simplifying assumptions, most of the free parameters in the model can be determined for each ion studied. The model is not intended to be a unique description of the data but to provide a plausible, semiquantitative description (see Marchais and Marty [1979] for a more quantitative treatment).

The shapes of the τ-V curves can be interpreted easily with the model described above. As the membrane is depolarized, the energy profile for a permeant ion will shift as shown in Figure 3. Since the internal barrier is much larger than the external barrier, the rate constant, c, will be rate-limiting, and the external ion will be in

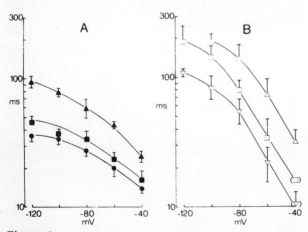

Figure 2
Averaged τ-V relations for monovalent and divalent cations. Averaged τ-V curves obtained at 12°C in the presence of Na^+ (●), Li^+ (■), Cs^+(▲), Ba^{++} (▽), Sr^{++} (□), and Mg^{++} (△) are compared. The curves are steeper for divalent ions (B) than for monovalent ions (A). (Reprinted, with permission, from Marchais and Marty 1979.)

equilibrium with the binding site. In this case, both the probability that the channel is occupied and the mean channel-open time will vary exponentially with voltage according to the Boltzmann distribution. This is seen for both monovalent and divalent ions in the τ-V curves shown in Figure 2. However, as the membrane is hyperpolarized, an energy profile more like that shown by the dotted line in Figure 3 will be obtained. When the two barriers are comparable in size, no one rate constant is limiting and the external ion can no longer be considered to be in equilibrium with the site. In this case, the mean channel-open time will depend less steeply on voltage than predicted by the Boltzmann equation. This saturation of τ at hyperpolarized voltages was found experimentally (Fig. 2).

ION INTERACTIONS CAN BE PREDICTED BY THE MODEL

• The permeation model described above assumes that only one ion at a time may occupy the site inside the channel. As a result, simple competition effects should be seen in solutions where more than one ion is present. Such a result is shown in Figure 4. Here, both the mean channel-open time and single-channel current have been measured by noise analysis in solutions in which varying proportions of Na^+ have been replaced isotonically with Ca^{++}. As might be expected from the results presented in Figure 2, Ca^{++} binds in the channel more strongly than Na^+ does, reducing the single-channel current and

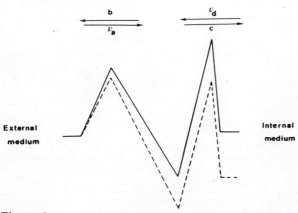

Figure 3
Energy profile in the channel. A simplified energy profile for Na^+ is illustrated at +20 mV (———) and −100 mV (--------). (Reprinted, with permission, from Marchais and Marty 1979.)

increasing the channel-open time. The points have been fit with the model described above. The model provides a good qualitative description of the data, though there are some numerical discrepancies. Notice that the model predicts correctly the nonlinear relationship between elementary current and $[Ca^{++}]_e$. If the two ions did not interact, but obeyed the independence principle, a straight line connecting the elementary current in 0% Ca^{++} (100% Na^+) to that in 100% Ca^{++} would be predicted. At intermediate $[Ca^{++}]_e$, the elementary current is smaller than predicted by a description in which the two ions permeate independently. The interpretation of the above model is that Ca^{++} is not only less permeable than Na^+, but also that it acts as a quasi-channel blocker, reducing the number of channels available for Na^+ permeation at any given instant.

GLUCOSAMINE—A BLOCKING ION

• It has been noted that the above model is similar formally to descriptions of channel blockers. In Figure 5, relaxation experiments in solutions containing the ion glucosamine are shown. Glucosamine seems to be a compound that is intermediate between a permeant ion and a conventional blocker. It might be expected that an ion that could enter the ACh channel and block it could be described by an

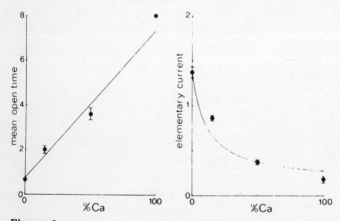

Figure 4
Mean channel-open time and elementary current in Na^+-Ca^{++} solutions. Mean channel-open times (*left*) and elementary currents (*right*) were compared in normal seawater (reference values not given) and in solutions containing essentially isotonic solutions of NaCl and $CaCl_2$ in various proportions. V_m, −80 mV. (Reprinted, with permission, from Marty 1980.)

energy profile similar to that shown in Figure 3, except that the internal barrier would be so large that virtually none of the blocking ions could pass over it into the cell. In this case, the external blocking ion would always be in equilibrium with the site in the channel, and no saturation of the τ-V curve would be seen even at very hyperpolarized voltages. This is the case for glucosamine: The exponential relationship between τ and V is maintained at the most-hyperpolarized voltages studied (Fig. 5B).

Figure 5A illustrates that the relaxation has a rather unusual form—the inward current actually decreases immediately after the start of the hyperpolarizing step. This initial decrease is due presumably to the fact that the sudden hyperpolarization drives glucosamine into channels that were open just prior to the application of the step, resulting in a rapid decrease in conductance. This rapid conductance change is followed by a slower redistribution of channels, which can be either blocked in their open state ($R_o \cdot I$), open and unblocked (R_o), or closed (R_c). The redistribution results in a greater mean number of open channels (and therefore an increased inward current) due to the glucosamine-induced slower channel-closing rate.

Another consequence of the interaction of glucosamine with the channel is an apparent voltage-sensitivity of the single-channel conductance (γ). V_m has two opposing effects on the single-channel current that is used to estimate γ. One effect is the increased driving force for permeant ions. The other is an increase in the fraction of time the channel is blocked by glucosamine. For values of V_m up to -70 mV, the single-channel current increases. Beyond -70 mV, the

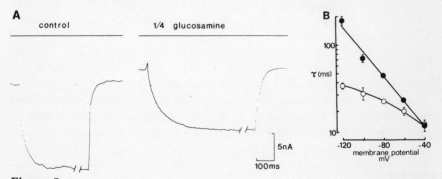

Figure 5
Effects of glucosamine on relaxation kinetics of ACh-induced currents in voltage-clamp experiments. (A) Relaxation experiment in control seawater (left) and after replacement of 25% of the Na^+ with glucosamine (right). Voltage jump, -60 mV to -100 mV. The background current has been subtracted. (B) Averaged τ values obtained from relaxation experiments in control seawater (o) and in 25% glucosamine seawater (•). (Reprinted, with permission, from Marchais and Marty 1980.)

188   Molluscan Nerve Cells

single-channel current declines. The low value found for the single-channel current in the presence of glucosamine presumably is due to very fast blocking and unblocking of open channels. These changes are too rapid to be resolved and result in a time-averaged lower value for the elementary current.

Adams and Sakmann (1978) used a description like equation 1 to describe the action of decamethonium, but they did not make the assumption that the second reaction was always in equilibrium, since they observed relaxations with two components. The rate constant for entry of glucosamine seems to be at least ten times faster than that for decamethonium, so the fast component of the relaxation cannot be resolved in this case (see Fig. 5A).

SUMMARY

• The mean open time of the ACh channel has been found to depend strongly on the species of permeant ion present and on voltage. The results can be explained simply if it is assumed that ions enter the channel, bind to a site within the membrane field, and, while in the site, prevent the channel from closing. Equation 1 may be used to describe the actions of permeant ions on channel lifetime, the effects of the blocking ion glucosamine, and, if the binding reaction is not assumed to be in equilibrium, the effects of conventional antagonists, such as tubocurarine and hexamethonium (Ascher et al. 1978b). In this description, all of the voltage dependence of the channel lifetime may be attributed to the voltage dependence of ion binding. This is in contrast to the descriptions of the ACh channel in frog end plate, in which the voltage dependence of the channel has been attributed either to dipole moments in the channel protein itself (Magelby and Stevens 1972b) or to voltage-dependent agonist binding (Kordaš 1969). Although it would be difficult to rule out completely any voltage dependence of α (the closing rate constant seen in the absence of permeant ions), attributing most of the voltage dependence to permeant-ion effects seems to be the simplest, least ad hoc way to describe the ACh channel in Aplysia.

REFERENCES

Adams, P.R. 1976a. Voltage dependence of agonist responses at voltage-clamped frog endplates. Pfluegers Arch. gesamte Physiol. Menschen Tiere 361: 145.

———. 1976b. Drug blockage of open end-plate channels. J. Physiol. (Lond.) 260: 531.

Adams, P.R. and B. Sakmann. 1978. Decamethonium both opens and blocks endplate channels. Proc. Natl. Acad. Sci. 75: 2994.

Anderson, C.R. and C.F. Stevens. 1973. Voltage clamp analysis of acetylcholine produced end-plate current fluctuations at frog neuromuscular junction. *J. Physiol. (Lond.)* **235:** 655.

Ascher, P., A. Marty, and T.O. Neild. 1978a. Life time and elementary conductance of the channels mediating the excitatory effects of acetylcholine in *Aplysia* neurones. *J. Physiol. (Lond.)* **278:** 177.

————. 1978b. The mode of action of antagonists of the excitatory response to acetylcholine in *Aplysia* neurones. *J. Physiol. (Lond.)* **278:** 207.

Fatt, P. and B. Katz. 1951. An analysis of the end-plate potential recorded with an intracellular electrode. *J. Physiol. (Lond.)* **115:** 320.

Kordaš, M. 1969. The effect of membrane polarization on the time course of the end-plate current in frog sartorius muscle. *J. Physiol. (Lond.)* **204:** 493.

Laeuger, P. 1973. Ion transport through pores: A rare-theory analysis. *Biochim. Biophys. Acta* **311:** 423.

Magleby, K.L. and C.F. Stevens. 1972a. The effect of voltage on the time course of end-plate currents. *J. Physiol. (Lond.)* **223:** 151.

————. 1972b. A quantitative description of end-plate currents. *J. Physiol. (Lond.).* **223:** 173.

Marchais, D. and A. Marty. 1979. Interaction of permeant ions with channels activated by acetylcholine in *Aplysia* neurones. *J. Physiol. (Lond.)* **297:** 9.

————. 1980. Action of glucosamine on acetylcholine-sensitive channels. *J. Membr. Biol.* **56:** 43.

Marty, A. 1980. Action of calcium ions on acetylcholine sensitive channels in *Aplysia* neurones. *J. Physiol. (Paris)* **76** (in press).

Marty, A., T.O. Neild, and P. Ascher. 1976. Voltage sensitivity of acetylcholine currents in *Aplysia* neurones in the presence of curare. *Nature* **261:** 501.

Neher, E. 1971. Two fast transient current components during voltage clamp on snail neurons. *J. Gen. Physiol.* **58:** 36.

Neher, E. and B. Sakmann. 1975. Voltage dependence of drug induced conductance in frog neuromuscular junction. *Proc. Natl. Acad. Sci.* **72:** 2140.

Neher, E. and C.F. Stevens. 1977. Conductance fluctuations and ionic pores in membranes. *Annu. Rev. Biophys. Bioeng.* **6:** 345.

Sheridan, R.E. and H.A. Lester. 1975. Relaxation measurements on the acetylcholine receptor. *Proc. Natl. Acad. Sci.* **72:** 3496.

Takeuchi, A. and N. Takeuchi. 1960. On the permeability of end-plate membrane during the action of transmitter. *J. Physiol. (Lond.)* **154:** 52.

Voltage-sensitive Postsynaptic Channels

Based on a presentation by

WILKIE A. WILSON*

Epilepsy Center
Veteran's Administration Hospital
Durham, North Carolina 27705

• Slow synaptic responses were long considered interesting but unusual phenomena of questionable physiological significance. However, synaptic responses lasting for several seconds or longer have now been described in many preparations, including vertebrates, and appear to play an important role in the control of neuronal activity. Recently, several studies have shown that some slow synaptic responses are not due to the opening of conventional voltage-independent synaptic channels (see Marty, this volume) but instead involve the modulation of voltage-sensitive conductances by the transmitter. Pellmar and Wilson (1977) provided the first evidence for voltage-sensitive synaptic conductances in a study of a slow excitatory response to serotonin applied iontophoretically to neurons in the mollusc *Aplysia*. Using voltage-clamp techniques, they found that serotonin produced a slow response with two components: (1) a conventional response present at all levels of V_m that appeared to be due to a voltage-independent increase in g_{Na}, and (2) a voltage-dependent response present at depolarized levels that was later shown to involve an increased g_{Ca} (Pellmar and Carpenter 1979).

Slow inhibitory synaptic responses can also involve voltage-sensitive channels. Two examples of slow inhibitory synaptic actions on voltage-sensitive channels are discussed here: a cholinergic response and a dopaminergic response in different identified neurons in the abdominal ganglion of *Aplysia*. These responses are particularly interesting because they both involve burst-firing neurons, and in each case the mechanism of slow synaptic inhibition is intimately related to the burst-generating mechanism.

*Work presented here was done in collaboration with S.M. Gospe.
The report of this presentation was prepared by E.T. Walters.

CHOLINERGIC INHIBITION OF LEFT-UPPER-QUADRANT BURSTING NEURONS

• The bursting neurons L2 and L6 in the left upper quadrant (LUQ) of the abdominal ganglion receive monosynaptic inhibitory connections from the identified cholinergic cell L10 (Frazier et al. 1967). In most of these cells, L10 produces both a brief (about 100 msec) unitary inhibitory PSP and a slower inhibitory response that lasts several seconds (Fig. 1A). These inhibitory responses were examined using voltage-clamp techniques to measure the IPSCs. Two electrodes were inserted into the postsynaptic cell for voltage clamping, and a third electrode was used to control the firing of L10. Figure 1B illustrates the IPSCs resulting from intracellular activation of the presynaptic cell L10 (the recording from L10 is not shown). At -30 mV, both the fast synaptic current and the slow synaptic current were outward. However, when the postsynaptic cell was hyperpolarized to -50 mV, the short IPSC disappeared; at more negative potentials it inverted, becoming inward. In contrast, the long IPSC approached zero at hyperpolarized potentials and never inverted despite prolonged high-frequency firing of L10. These data are summarized in the I-V curves of Figure 1C. The short IPSC inverts at E_{Cl}, suggesting that this component of the inhibitory response is mediated by Cl^- (Blankenship et al. 1971).

SLOW PSP IS DUE TO A DECREASE IN A VOLTAGE-DEPENDENT INWARD CURRENT

• Because the long IPSPs can modify the bursting rhythms of these cells, Wilson and Wachtel (1978) examined the possibility that the long IPSC might be related to the intrinsic burst generator.

Wilson and Wachtel (1974) had found previously that in the normal I-V plots of burst-firing neurons, there is a region of negative slope (see Fig. 2B) in the unstable voltage range through which V_m oscillates. This negative-resistance characteristic was produced by a slow regenerative inward current (mediated by Na^+, Ca^{++}, or both) that appeared to be responsible for the oscillations. Therefore, Wilson and Wachtel (1978) examined the effect of the long IPSP on the negative-resistance region of the I-V curves of these bursting neurons. With L10 held silent, a hyperpolarizing voltage command was given to the follower cell (Fig. 2A). This resulted in an incremental outward current, which indicated the presence of a negative-resistance characteristic. The cell was then returned to V_h, and L10 was fired. During the resulting prolonged outward IPSC, a second hyperpolarizing command was given to the follower cell; in this case, the response was an

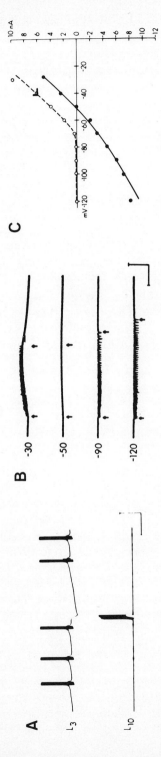

Figure 1
Inhibition of LUQ cells by interneuron L10. (A) Firing L10 causes a long-lasting hyperpolarization of L3 concomitant with an interruption of bursting. (B) Membrane current recorded from L3 during firing of L10 (recording from L10 not shown). As L3 is clamped at progressively more-hyperpolarized levels, the fast IPSC inverts from an outward to an inward current and the slow IPSC decreases to zero without inverting. (C) I-V curves of the fast (●) and slow (○) components of the IPSC elicited in L3 by firing L10. The short component inverts near E_{Cl}. (Reprinted, with permission, from Pellmar and Wilson 1977.)

incremental inward current, which indicated that the negative-resistance characteristic had been lost. Repeated use of this procedure
over a range of voltage commands yielded *I-V* curves for the cell in
the presence and absence of the prolonged IPSC (Fig. 2B). The curve
obtained in the absence of the IPSC showed the typical negative-
resistance characteristic described by Wilson and Wachtel (1974).
However, the curve obtained during the long IPSC lacked this negative-resistance region. Instead, the *I-V* curve had a positive, nearly
constant slope. Therefore, synaptic input appears to occlude the
negative-resistance region, transforming a burst-firing neuron into an
ordinary neuron.

This conclusion was supported by experiments in which ACh
was iontophoresed onto the axons of LUQ bursting neurons. Direct
application of ACh produced prolonged outward currents, which,
like the long IPSC, did not invert with hyperpolarization. Moreover,
this response to ACh also occluded the negative-resistance region. In
both cases of occlusion, there was no increased conductance at voltages outside of the negative-resistance region. This argues against the
mechanism being conventional increased g_K, since such a mechanism
would be expected to extend to voltages beyond the negative-resistance region. Further support for this conclusion came from experiments showing that increased $[K^+]_e$ had no effect on the response. On
the other hand, replacing Na^+ in the bath eliminated the inhibitory
response. Thus, the ionic requirements of this response parallel those

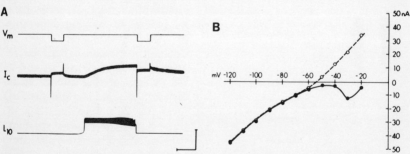

Figure 2
Synaptic action of L10 occludes the negative-resistance characteristic of L3.
(*A*) L3 was clamped at −35 mV, and 10-mV hyperpolarizing commands were
given before and then after firing interneuron L10. The first command resulted in an outward current, which indicated that L3 has a negative-
resistance characteristic. The second command, given during the slow IPSC,
resulted in an inward current, which indicated that the negative resistance
had been lost. (*B*) *I-V* plots for L3 before prolonged inhibition (•) and during
the long IPSC (o). The long IPSC eliminates the negative-resistance region
seen prior to firing L10. (Reprinted, with permission, from Pellmar and
Wilson 1977.)

of burst generation and suggest that the same slow inward current may be involved in both.

DOPAMINERGIC INHIBITION OF R15

• R15 is an identified neuroendocrine cell in the abdominal ganglion of *Aplysia* that has long been known for its distinctive, regular bursting pattern (Arvanitaki and Chalazonitis 1958). Occasionally, R15 receives slow inhibitory synaptic input, which, like the slow IPSC of the LUQ bursting neurons (see above), cannot be inverted by hyperpolarization. Although the source of this inhibitory input to R15 is unknown, the effects can be mimicked by introducing dopamine (50–500 μM) into the bath. Figure 3 illustrates the hyperpolarization of R15 and the elimination of spontaneous bursting by dopamine. Since repolarization of the neuron by injection of depolarizing current could not induce bursting, it appears that the hyperpolarization itself is not responsible for the inhibition of bursting. Bursting reappeared when the dopamine was washed out. By using voltage-clamp techniques similar to those used to examine the slow cholinergic inhibition of the LUQ bursting neurons, Wilson and Wachtel (1978) and Gospe and Wilson (1980) found that before application of

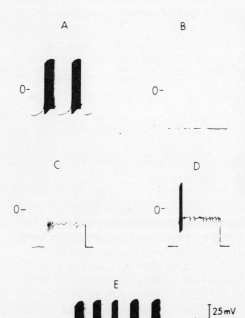

Figure 3
Dopamine hyperpolarizes R15 and abolishes bursting. (*A*) Burst activity before introducing dopamine. (*B*) After introduction of 500 μM dopamine, the cell hyperpolarizes to a stable V_R. (*C, D*) 15 nA and 25 nA of depolarizing current to return R15 to near its original resting level fail to induce bursting. (*E*) Bursting reappears after washing. (Reprinted, with permission, from Gospe and Wilson 1980.)

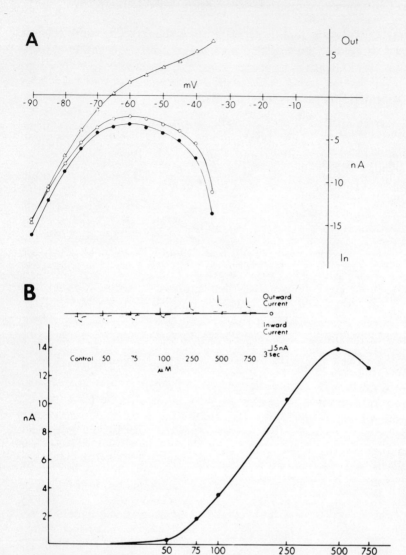

Figure 4
Dopamine eliminates the negative-resistance characteristic in R15 in a
dose-dependent manner. (*A*) *I-V* curve obtained for R15 under voltage-
clamp conditions (●). Elimination of negative-resistance region in *I-V*
curve by 500 μM dopamine (Δ). The effect reverses with washing (○). (*B*)
A dose-response curve of the effect of dopamine on the current at a
given voltage. (*Inset*) Current responses caused by stepping V_m from V_h
of -60 mV to -35 mV at different dopamine concentrations. The curve
is a measure of the difference in current from the control value and
shows the progressive loss of inward current with increasing concentrations of dopamine. (Reprinted, with permission, from Gospe and
Wilson 1980.)

dopamine, R15 has a region of negative resistance in its I-V curve (Fig. 4A). However, during the prolonged inhibition produced by dopamine, the negative-resistance region is eliminated, with little alteration of the I-V curve in the region more negative than -75 mV. After the dopamine is washed out, the region of negative resistance returns. As is found with the slow cholinergic inhibition of the LUQ bursting neurons, the slow dopaminergic effect on R15 was unaffected by changes in $[K^+]_e$ but was eliminated by removing external Na^+. Thus, dopamine appears to inhibit the regenerative inward current responsible for the negative-resistance region and burst generation.

What is the mechanism by which transmitter occludes negative-resistance characteristics of bursting cells? A first step in answering this question is to elucidate the pharmacology of the effect. The dopamine response in R15 is advantageous for pharmacological investigations because, unlike many transmitter effects, this prolonged response shows very little desensitization. This allows the acquisition of detailed dose-response curves for individual cells. Consecutive doses of dopamine in increasing concentrations were applied to R15. An I-V curve was obtained at each concentration. Figure 4B shows the effect of dopamine, at six different concentrations, on the amount of current flowing when V_m was clamped to -35 mV (in the negative-resistance region). With increasing doses of dopamine, the occlusion of inward current increased progressively in a sigmoidal fashion and rose from minimum to maximum as drug concentration increased by one order of magnitude. This information was used to construct a standard dose-response curve, which is useful because it allows precise comparison of various agonists and antagonists and may make it possible to find selective antagonists that could simplify the analysis of the complex slow inhibitory effect on R15.

CONCLUSION

• The ultimate goal of these investigations is to determine the intra-cellular mechanisms responsible for slow synaptic responses. It is interesting that synaptic activity and axonal application of transmitter can eliminate the negative-resistance characteristic in R15, a cell whose isolated soma can exhibit bursting (Alving 1968). One implication of these observations, consistent with the slow time course of the inhibitory response and with observations of other slow transmitter effects in *Aplysia* (Klein and Kandel 1978), is that an intracellular second messenger could mediate the occlusive effect on the burst-generating slow inward current.

REFERENCES

Alving, B.O. 1968. Spontaneous activity in isolated somata of *Aplysia* pacemaker neurons. *J. Gen. Physiol.* **51:** 29.

Arvanitaki, A. and N. Chalazonitis. 1958. Configurations modales de l'activité, progres à différents neuronas d'un même centre. *J. Physiol. (Paris)* **50:** 122.

Blankenship, J.E., H. Wachtel, and E.R. Kandel. 1971. Ionic mechanisms of excitatory, inhibitory and dual synaptic actions mediated by an identified interneuron in abdominal ganglion of *Aplysia*. *J. Neurophysiol.* **34:** 76.

Frazier, W.T., E.R. Kandel, I. Kupfermann, R. Waziri, and R.E. Coggeshall. 1967. Morphological and functional properties of identified neurons in the abdominal ganglion of *Aplysia californica*. *J. Neurophysiol.* **30:** 126.

Gospe, S.M. and W.A. Wilson. 1980. Dopamine inhibits burst firing of neurosecretory cell R15 in *Aplysia californica*: Establishment of a dose response relationship. *J. Pharmacol. Exp. Ther.* **214:** 112.

Klein, M. and E.R. Kandel. 1978. Presynaptic modulation of voltage-dependent Ca^{2+} current: Mechanism for behavioral sensitization in *Aplysia californica. Proc. Natl. Acad. Sci.* **75:** 3512.

Pellmar, T.C. and D.O. Carpenter. 1979. Voltage-dependent calcium current induced by serotonin. *Nature* **277:** 483.

Pellmar, T.C. and W.A. Wilson. 1977. Unconventional serotonergic excitation in *Aplysia. Nature* **269:** 76.

Wilson, W.A. and H. Wachtel. 1974. Negative resistance characteristic essential for the maintenance of slow oscillations in bursting neurons. *Science* **186:** 932.

————. 1978. Prolonged inhibition in burst firing neurons: Synaptic inactivation of the slow regenerative inward current. *Science* **202:** 772.

Modulation of Presynaptic Ca^{++} Currents and the Control of Transmitter Release

Presented by

ELI SHAPIRO*

Division of Neurobiology and Behavior
Department of Physiology
College of Physicians & Surgeons
of Columbia University
New York, New York 10032

● The diversity of voltage-independent ionic conductance channels has been examined earlier in this volume. With this background, it becomes possible to examine how these channels are modulated. In particular, one may examine the functional relationships between the regulation of currents underlying the membrane and action potential in the presynaptic terminals and the regulation of transmitter release. A review of this type of analysis based on studies of the squid giant synapse is given by Llinás (this volume). However, whereas output of transmitter from the squid synapse is relatively stable, many synapses in the central nervous system undergo marked changes in their ability to release transmitter, both as a function of use or as the result of various extrinsic factors. This property of synapses to change their effectiveness is known as plasticity. In central synapses that show plasticity, it is possible, therefore, to extend the work of Llinás and his colleagues on the relationship between I_{Ca} and transmitter release (Llinás, this volume) by examining the ionic mechanisms that underlie various forms of synaptic plasticity. Specifically, one can examine whether synaptic plasticity is due either to modulation of the presynaptic I_{Ca} or to modulation of the coupling between presynaptic I_{Ca} and transmitter release.

The exploration of such questions requires a system such as the squid giant synapse in which the presynaptic terminal and postsynaptic cells are readily accessible for electrical recording and voltage-

*Work presented here was done in collaboration with V. Castellucci and E. Kandel and supported by NIH grants NS12744 and GM23540, Career Scientist Award MH18558 to E. Kandel, and NIH postdoctoral fellowship 2F32-NS05393 to E. Shapiro.

clamping. In addition, the system should be capable of being modulated by intrinsic or extrinsic factors.

We have begun to explore synaptic modulation in a system that meets these requirements: the excitatory and inhibitory synapses made by cholinergic interneuron L10 onto identified follower cells (left-upper-quadrant cells [LUQC] and cells of the RB cluster) in the abdominal ganglion of *Aplysia californica* (Kandel et al. 1967). Like other molluscan somata, the soma of L10, the presynaptic neuron, has Ca^{++} channels (Shapiro et al. 1980a). In addition, for another *Aplysia* neuron, it has been shown recently that changes in the Ca^{++} current in the soma parallel changes in transmitter release (Klein, this volume). These two observations suggested that combining voltage-clamp of the soma of the presynaptic neuron with recording of postsynaptic responses would make it possible to correlate the presynaptic I_{Ca} with transmitter release.

Transmitter release at the synapse made by L10 undergoes two types of modulation. First, spike-elicited release can be varied by changing V_R of the soma (intrinsic modulation). This type of synaptic modulation, which has been described previously at other central synapses (Shimahara and Tauc 1975; Waziri 1977; Nicholls and Wallace 1978a; Shimahara and Peretz 1978), suggests that the transmitter-release sites of L10 are electrically close to the soma. Second, cell L10 also undergoes presynaptic inhibition, a form of extrinsic modulation (Waziri et al. 1969; Shapiro et al. 1980b). In vertebrates, presynaptic inhibition has been correlated with presynaptic depolarization (reviewed by Burke and Rudomin 1977; Ryall 1978). However, since depolarization of L10 increases transmitter release, it is likely that other mechanisms may be involved in the presynaptic inhibition at these synapses.

The work reviewed here (see also Shapiro et al. 1980a,b) was designed to correlate I_{Ca} with transmitter release and then to examine I_{Ca} during two types of synaptic modulation: (1) control of transmitter release by presynaptic V_m and (2) presynaptic inhibition.

PROPERTIES OF Ca^{++} CURRENT AND RELEASE

• By combining a voltage-clamp analysis of L10 with standard pharmacological separation techniques, the voltage sensitivities of transmitter release and presynaptic I_{Ca} were compared. These experiments also allowed the evaluation of the electrical control of transmitter-release sites by means of electrodes placed in the soma of L10.

To maximize voltage-clamp control, L10 was axotomized at its exit from the abdominal ganglion to eliminate uncontrolled axon currents (Connor 1977) and the preparation was treated with TTX to block regenerative I_{Na}. The first criterion for a well-controlled cell was the elimination of axon currents (A spikes). Under these conditions, depolarizing voltage-clamp steps continue to elicit PSPs in about 50% of cases. Within a range of V_m from about −35 mV (the threshold for large, inward I_{Ca}) to about +10 mV, both I_{Ca} in L10 and transmitter release (as monitored by the size of the PSP in the follower cells) increase with increasing depolarization (Fig. 1A). Within this V_m range, there is a linear relationship between presynaptic I_{Ca} and postsynaptic response (Fig. 1B). In the range of clamp duration from 10−40 msec, increasing the duration of the depolarizing clamp step also causes increased transmitter release (Fig. 1C).

These results agree with the transfer characteristics described by Llinás and his coworkers (Llinás et al. 1976) for transmitter release at the squid giant synapse. In the case of L10, however, synaptic V_m cannot be controlled through as wide a range as at the squid terminal.

The finding that transmitter release can be a graded function of both voltage-step amplitude and duration indicates that the synaptic terminal membrane can be controlled, at least partially, with electrodes placed in the soma of L10. This preparation therefore seemed amenable to an analysis of the mechanisms underlying synaptic modulation. Small voltage steps, in the range of voltages where control appeared optimal (−35 mV to 0 mV), were used.

CONTROL OF TRANSMITTER RELEASE BY PRESYNAPTIC V_R

• V_R has a profound effect on transmitter release in response to a transient depolarization. When L10 is held at −40-mV, action potentials elicit large PSPs. As the V_R of L10 is hyperpolarized, action-potential-evoked transmitter output is decreased.

In L10, spikes elicited from a depolarized V_m are broader than spikes from hyperpolarized levels. The fact that the duration of depolarization can affect transmitter release (Fig. 1C) suggests that the change in spike configuration may play a role in synaptic modulation (Katz and Miledi 1965; Klein, this volume).

To examine properties of the ionic currents of L10 that might underlie this type of modulation, L10 was voltage-clamped to either a depolarized or hyperpolarized V_h. Step depolarizations to the same V_m elicited different-sized PSPs from two different holding levels (Fig. 2A). Clamp currents elicited from hyperpolarized V_h levels invariably are more outward than currents elicited from depolarized levels (Fig. 2A). This difference is due to the removal of inactivation of the two voltage-sensitive K^+ channels, I_A and $I_{K(V)}$ (Connor; Thompson; both this volume). These changes in K^+ currents would control the duration of spikes elicited from different V_R levels. Therefore, one

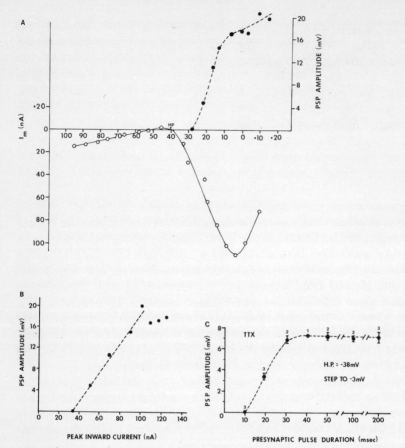

Figure 1

Transmitter release and presynaptic I_{Ca}. (A) Cell L10 was voltage-clamped at a V_h of -40 mV. Transmitter release was evoked by 200-msec-duration depolarizing clamp steps. The size of the PSP (•) recorded in cell L5 (LUQC) was plotted against step depolarization. The preparation was then treated with 4-AP (10 mM) and TEA (25 mM) to block voltage-activated I_K; extracellular Ca^{++} was replaced by Ba^{++} to block $I_{K(Ca)}$. The peak inward current (o) evoked by 200-msec step depolarizations from -40 mV to various voltages is plotted in the lower part of the figure. (B) Plot of the relationship between peak inward current and PSP amplitude for this cell. Inward current was leakage-corrected. (C) PSP amplitude (cell L5) is a graded function of duration of presynaptic (L10) depolarization. Depolarizing voltage-clamp steps to -3 mV from V_h of -38 mV were delivered once every 30 sec. Duration of this pulse was varied. From durations of 10 msec to about 40 msec, the size of the PSP is a graded function of the presynaptic depolarization. The number of PSPs, averaged for each duration, is given above each point. Responses were recorded in high-divalent cation seawater (60 mM Ca^{++}, 140 mM Mg^{++}, 265 mM Na^+) containing TTX.

202

mechanism for the modulation of transmitter release by V_R is by means of a modulation, through control of spike duration, of the amount of Ca^{++} entering during a spike. Since the duration of depolarization is held constant in the voltage-clamp experiments (Fig. 2A, B), however, mechanisms in addition to changes in spike duration may be contributing to the altered transmitter release at different holding potentials.

I_{CA} MODULATION BY V_R

• Examination of the voltage-sensitive properties of the Ca^{++} channel suggests another mechanism for the modulation of transmitter release by changes in V_R. The transient I_{Ca} elicited by voltage-clamp steps is relatively unaffected by the value of V_h (Fig. 2B, C). If anything, the transient I_{Ca} elicited by steps from depolarized levels may be smaller than that evoked from hyperpolarized levels (Fig. 2C). However, as first described in *Helix* neurons (Eckert and Lux 1976; Brown, this volume), the Ca^{++} channel also has a steady-state activation property (Fig. 2D; see also Fig. 2E). Examination of steady-state *I-V* relationships in L10 showed that in the range of V_m around normal V_R (-40 mV to -60 mV), depolarization activates a steady, slow I_{Ca}.

The demonstration of a steady-state I_{Ca} suggests that in addition to affecting spike duration, V_m can also modulate transmitter release in a second way. The steady-state I_{Ca} may increase the steady-state $[Ca^{++}]_i$, and thus increase the ability of a relatively constant transient I_{Ca} to cause transmitter release.

This interpretation of a steady-state I_{Ca} activated when V_R is relatively positive is consistent with the results of Nicholls and Wallace (1978b) on leech neurons. They found increases in quantal PSP frequency at depolarized values of presynaptic V_m.

Thus, control of transmitter release by presynaptic V_m involves a direct modulation of the Ca^{++} channel, allowing changes in the coupling between transient Ca^{++} influx and transmitter release as well as the indirect modulation of the transient Ca^{++} influx by spike duration.

PRESYNAPTIC INHIBITION

• Presynaptic inhibition is a heterosynaptic, or extrinsic, form of synaptic modulation in which stimulation of a modulating pathway causes a decrease of transmitter output. A brief train of electrical stimuli delivered to the connectives that link the abdominal ganglion with the head ganglia inhibits, presynaptically, the PSPs produced by

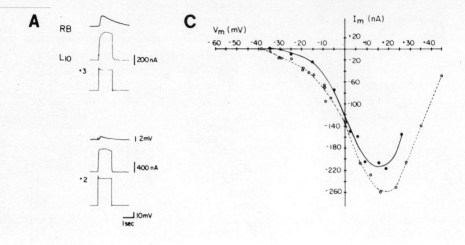

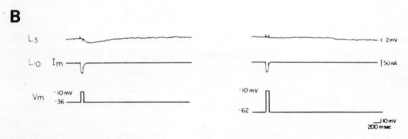

Figure 2

(A) Presynaptic cell L10 is voltage-clamped at a V_h of -45 mV (*top set*) and -62 mV (*bottom set*) in an artificial seawater solution containing 30 μM TTX. Step depolarizations of 1.0 sec from each V_h (interpulse interval = 1 min) evoke EPSPs in the follower cell (top trace in each set). The middle trace in each set is the monitored current from L10 and the bottom trace is the voltage record. From a V_h of -62 mV, step depolarizations elicit smaller PSPs. Each clamp current is more outward than for steps from -45 mV. (B) Amplitude of PSP is still modulated by V_h after blockage of most I_K. Presynaptic neuron L10 is voltage-clamped from two levels of V_h in a Ca^{++}-free solution containing 60 mM Ba^{++}, 30 μM TTX, 25 mM TEA, and 10 mM 4-AP. A 100-msec step depolarization to -10 mV from a V_h of -36 mV elicits a large transient inward current and a large IPSP. A 100-msec step depolarization to -10 mV from a V_h of -62 mV elicits a transient inward current as large as from -36 mV but only a small IPSP. (C) Transient I_{Ca} is independent of V_h. The I-V relationship is shown for I_{Ca} of the presynaptic neuron L10 from two levels of V_h: -60 mV (o---o) and -40 mV (•——•). The experiment was carried out in a Na^+-free solution containing 265 mM TEA and 5 mM 4-AP. The peak transient inward current is not increased when elicited from depolarized levels of V_h. It may be slightly decreased, as is the case here, perhaps due to steady-state inactivation or build-up of $[Ca^{++}]_i$. The curves are leakage-corrected.

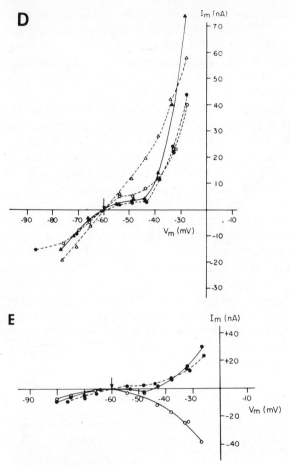

D

E

Figure 2 (*continued*)

(*D, E*) Steady-state *I-V* curves of L10. In *D* and *E*, L10 was voltage-clamped to a V_h of −60 mV, and 5-sec clamp steps were delivered every 30 sec to various hyperpolarized and depolarized V_ms. The current at the end of each 5-sec voltage step was plotted against V_m. In normal seawater with TTX (○ in *D*; ●---● in *E*), the steady-state *I-V* curve shows a pronounced region of reduced positive slope in the range of the normal V_R of L10 (−60 mV to −40 mV). In *D*, application of TEA and 4-AP (●) have little effect on the steady-state *I-V* curve, since transient K^+ channels normally inactivate during the prolonged (5-sec) voltage steps. Replacement of extracellular Ca^{++} with 10 mM Co^{++} (Δ), however, linearizes the *I-V* curve by blocking an inward steady-state I_{Ca}. When normal seawater is washed back (▲), the normal steady-state *I-V* curve is restored. In *E*, when Ba^{++} (○) is substituted for Ca^{++}, the normally observed area of steady-state reduced positive slope is converted into a region of inward-going rectification. Ba^{++} carries the steady-state inward current through the Ca^{++} channel but does not activate an opposing $I_{K(Ca)}$. (●——●) Normal seawater wash. TTX was present in all solutions. (*C* is reprinted, with permission, from Shapiro et al. 1980a.)

L10 in its follower cells. Such stimuli reduce for approximately 30 sec the amplitude of the PSPs initiated by L10 without any significant alteration of the input resistance of the postsynaptic cells, suggesting that the reduction of the PSP is due to a presynaptic mechanism.

In *Aplysia*, presynaptic inhibition usually is correlated with presynaptic hyperpolarization (Waziri et al. 1969). However, the inhibitory effect outlasts the hyperpolarization and also can be seen in the rare cases when L10 is depolarized. Voltage-clamp experiments revealed that presynaptic inhibition could be elicited even when PSPs are elicited by voltage-clamp polarizations (Fig. 3A). Presynaptic inhibition invariably makes less inward the clamp currents that flow during depolarizing commands when either steady-state (Fig. 3B) or transient I_{Ca} (Fig. 3C) is examined in isolation. The presynaptic conductance change during presynaptic inhibition has the same voltage sensitivity as does the Ca^{++} channel (Fig. 3C).

Thus, presynaptic inhibition in *Aplysia* seems to be caused by the transmitter-mediated decrease of presynaptic I_{Ca}. These results demonstrate a third form of presynaptic modulation—a direct decrease of the transient I_{Ca} and steady-state I_{Ca} caused by a neurotransmitter. Similar results have been reported in a number of vertebrate preparations in which various neurotransmitters have been shown to modulate voltage-sensitive Ca^{++} channels (Dunlap and Fischbach 1978; Mudge et al. 1979; Horn and McAfee 1980).

DISCUSSION

L10 Is a Useful Model of the Presynaptic Membrane

• The cell body of L10 is a useful model of a presynaptic membrane for several reasons. First, it is possible to insert two large electrodes into the presynaptic neuron for voltage-clamping. Second, the synapses of L10 are plastic and, therefore, allow experiments to be performed that are not possible at the squid giant synapse. Third, it is possible to control transmitter release from the cell body.

Work on L10 leads to the conclusion that a number of types of synaptic modulation result from the modulation of presynaptic I_{Ca}. We have found that I_{Ca} can be modulated in three ways: (1) indirectly by modulation of opposing K^+ currents, which control spike amplitude and duration; (2) by direct effects of V_m on steady-state Ca^{++} current; and (3) by direct action on I_{Ca} by a modulating neurotransmitter.

These conclusions are based on two critical assumptions:

1. It is assumed that the electrical properties of soma Ca^{++} channels are similar to those of Ca^{++} channels found in

A

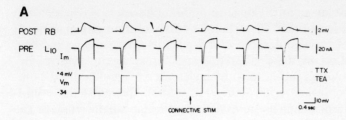

POST RB

PRE L10

I_m

+4 mV
V_m
-34

CONNECTIVE STIM

2 mV

20 nA

TTX
TEA

10 mV
0.4 sec

B

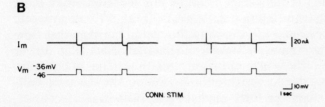

I_m

V_m -36 mV
-46

CONN. STIM.

20 nA

10 mV
1 sec

C

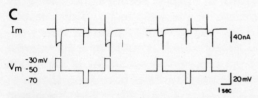

I_m

V_m -30 mV
-50
-70

40 nA

20 mV
1 sec

Figure 3
(A) Presynaptic inhibition of cell L10 under voltage-clamp control. In normal seawater containing TTX (30 μM) and TEA (50 mM) and with the cell held at -34 mV to inactivate I_A, a step depolarization elicits in the presynaptic neuron an inward I_{Ca} and an outward current due to $I_{K(Ca)}$. The depolarizing step also elicits an EPSP in the RB follower cell. After stimulation of the presynaptic inhibitory pathway for 5 sec, the same step depolarization elicits a decreased inward I_{Ca} and a reduced PSP. (B) Stimulation of a presynaptic inhibitory pathway reduces steady-state inward currents. Cell L10 is voltage-clamped in high-Ba^{++} seawater containing TTX (30 μM), TEA (25 mM), and 4-AP (10 mM). A small depolarizing voltage step produces a steady-state inward current. Stimulation of the presynaptic inhibitory pathway (for 5 sec) reduces this inward current. (C) The conductance changes caused by stimulation of the presynaptic inhibitory pathway are voltage-sensitive. Cell L10 is voltage-clamped in seawater containing Ba^{++}, TTX (30 μM), TEA (25 mM), and 4-AP (10 mM). Under these pharmacological conditions, 30-min intervals were used between runs. From a V_h of -50 mV, alternating 20-mV depolarizing and hyperpolarizing voltage steps elicit inward currents. Connective stimulation markedly reduces inward I_{Ba} through the Ca^{++} channel during the depolarizing step and only slightly increases "leakage" current of the hyperpolarizing step. (A and C are reprinted, with permission, from Shapiro et al. 1980b.)

the terminal region. This assumption appears to be reasonable for several reasons. The properties of Ca^{++} channels in the cell body of L10 resemble those of the squid synapse. In addition, changes in soma Ca^{++} spikes parallel changes in transmitter release in *Aplysia*. Finally, when L10 is voltage-clamped, there is good correlation between soma I_{Ca} and transmitter release.

2. It is assumed that the partial control of terminal voltage from the soma of L10 is sufficient to support the interpretation of data obtained with voltage clamp. This assumption is supported by the finding that cells treated with TTX and axotomized to prevent axon currents did not show A spikes but did show Ca^{++} currents similar to those recorded from intact cells under voltage clamp. In addition, in a range of V_m from −50 mV to +10 mV, variable amplitude and duration voltage-clamp commands produce graded PSPs. If terminals were not under control, one would expect all-or-none synaptic responses.

Conceptual Model of Presynaptic Terminals and Ca⁺⁺ Channels

The results presented here indicate that I_{Ca} can be modulated extrinsically by chemical transmitters and intrinsically and extrinsically by V_m. In addition, in a range of V_m from −50 mV to +10 mV, passing variable amplitude or duration depolarizing commands into the soma elicits graded PSPs.

These results and those of Klein and Kandel (1978; Klein, this volume) and Reuter and Scholz (1977) have led us to develop a conceptual model of the Ca^{++} channel (Fig. 4A, B). This intrinsic membrane protein has a number of putative sites of modulation. We have explored the role of V_R in the modulation of I_{Ca} and transmitter release (V_m gate, Fig. 4B). We also have examined the ability of the Ca^{++} channel to be modulated by a neurotransmitter (transmitter receptor gate, Fig. 4B). This modulation either may be direct or may be mediated by a second messenger. Work on cardiac muscle (Reuter and Scholz 1977; Tsien 1977; Reuter 1979) has provided evidence that transmitter modulation of Ca^{++} channels in this tissue can be modulated by cyclic nucleotides (cAMP gate, Fig. 4B). The Ca^{++} channel also can be modulated by $[Ca^{++}]_i$ (Brehm and Eckert 1978; Tillotson 1979 and this volume) (Ca^{++} gate, Fig. 4B).

These findings make the Ca^{++} channel a likely site for both long- and short-term mechanisms of synaptic modulation and provide a model whereby many cellular, metabolic, and hormonal processes can affect synaptic transfer functions.

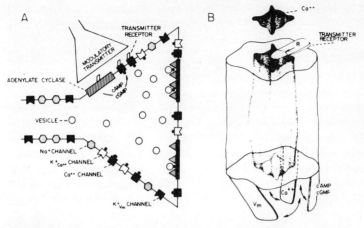

Figure 4
Models of modulation of I_{Ca}. (A) Conceptual model of synaptic
terminal-release area, illustrating several intrinsic membrane
proteins, the voltage-dependent Na$^+$ channels, the voltage-dependent K$^+$ channels, the Ca^{++} channels, and the Ca^{++}-depen-
dent K$^+$ channels. I_{Ca} can be modulated indirectly by increases
or decreases in I_K and by a direct action on the Ca^{++}-channel
protein (see B). (B) General model of the Ca^{++} channel. The
channel can be modulated by four regulatory signals: (1) V_m, (2)
[Ca^{++}]$_i$, (3) second messenger, and (4) modulatory transmitters
(see text for details). The channel has a selectivity filter that may
be a Ca^{++}-recognition site. (Reprinted, with permission, from
Shapiro et al. 1980b.)

REFERENCES

Brehm, P. and R. Eckert. 1978. Calcium entry leads to inactivation of calcium
   channel in *Paramecium*. *Science* **202:** 1203.
Burke, R.E. and R. Rudomin. 1977. Spinal neurons and synapses. In *Hand-
   book of physiology* (ed. E.R. Kandel), vol. 1, part 2, p. 877. American
   Physiological Society, Bethesda, Maryland.
Connor, J. 1977. Time course separation of two inward currents in molluscan
   neurons. *Brain Res.* **119:** 487.
Dunlap, K. and G.D. Fischbach. 1978. Neurotransmitters decrease the calcium
   component of sensory neurone action potentials. *Nature* **276:** 837.
Eckert, R. and D. Lux. 1976. A voltage-sensitive persistent calcium conductance in neural somata of *Helix*. *J. Physiol. (Lond.)* **254:** 129.
Horn, J. and D. McAfee. 1980. Alpha-adrenergic inhibition of calcium-dependent potentials in rat sympathetic neurones. *J. Physiol. (Lond.)*
   301: 191.
Kandel, E.R., W.T. Frazier, R. Waziri, and R.E. Coggeshall. 1967. Direct and
   common connections among identified neurons in *Aplysia*. *J. Neurophysiol.* **30:** 1352.

Katz, B. and R. Miledi. 1965. Release of acetylcholine from a nerve terminal by electric pulses of variable strength and duration. *Nature* **207**: 1097.

Klein, M. and E.R. Kandel. 1978. Presynaptic modulation of voltage-dependent Ca^{++} current: Mechanism for behavioral sensitization in *Aplysia californica*. *Proc. Natl. Acad. Sci.* **75**: 3512.

Llinás, R., I. Steinberg, and K. Walton. 1976. Presynaptic calcium currents and their relation to synaptic transmission: Voltage clamp study in squid giant synapse and theoretical model for the calcium gate. *Proc. Natl. Acad. Sci.* **73**: 2918.

Mudge, A.W., S.E. Leeman, and G.D. Fischbach. 1979. Enkephalin inhibits release of substance P from sensory neurons in culture and decreases action potential duration. *Proc. Natl. Acad. Sci.* **76**: 526.

Nicholls, J. and B.G. Wallace. 1978a. Modulation of transmission at an inhibitory synapse in the central nervous system of the leech. *J. Physiol. (Lond.)* **281**: 157.

————. 1978b. Quantal analysis of transmitter release at an inhibitory synapse in the central nervous system of the leech. *J. Physiol. (Lond.)* **281**: 171.

Reuter, H. 1979. Properties of two inward membrane currents in the heart. *Annu. Rev. Physiol.* **41**: 413.

Reuter, H. and H. Scholz. 1977. The regulation of the calcium conductance of cardiac muscle by adrenaline. *J. Physiol. (Lond.)* **264**: 49.

Ryall, R.W. 1978. Presynaptic inhibition. *Trends Neurosci.* **1**: 164.

Shapiro, E., V.F. Castellucci, and E.R. Kandel. 1980a. Presynaptic membrane potential affects transmitter release in an identified neuron in *Aplysia* by modulating the Ca^{2+} and the K^+ currents. *Proc. Natl. Acad. Sci.* **77**: 629.

————. 1980b. Presynaptic inhibition in *Aplysia* involves a decrease in the Ca^{2+} current of the presynaptic neuron. *Proc. Natl. Acad. Sci.* **77**: 1185.

Shimahara, T. and B. Peretz. 1978. Soma potential of an interneurone controls transmitter release in a monosynaptic pathway in *Aplysia*. *Nature* **273**: 158.

Shimahara, T. and L. Tauc. 1975. Multiple interneuronal afferents to the giant cells in *Aplysia*. *J. Physiol. (Lond.)* **247**: 299.

Tillotson, D. 1979. Inactivation of Ca conductance dependent on entry of Ca ions in molluscan neurons. *Proc. Natl. Acad. Sci.* **76**: 1497.

Tsien, R.W. 1977. Cylic AMP and contractile activity in the heart. *Adv. Cyclic Nucleotide Res.* **8**: 363.

Waziri, R. 1977. Presynaptic electrical coupling in *Aplysia*: Effects on postsynaptic chemical transmission. *Science* **195**: 790.

Waziri, R., E.R. Kandel, and W.T. Frazier. 1969. Organization of inhibition in abdominal ganglion of *Aplysia*. II. Post-tetanic potentiation, heterosynaptic depression, and increments in frequency of inhibitory postsynaptic potentials. *J. Neurophysiol.* **32**: 509.

Modulation of Ionic Currents and Regulation of Ca^{++} Influx during Habituation and Sensitization in *Aplysia*

Presented by

MARC KLEIN*

Division of Neurobiology and Behavior
Department of Physiology
College of Physicians & Surgeons
of Columbia University
New York, New York 10032

• The function of ionic currents in nerve cells is to participate in the transfer of information between those cells and other nerve cells or between those cells and receptor or effector end organs. In most cases, the ultimate result of such transfer of information is the initiation or modulation of behavior. Many behaviors can be modified by experience, and in some cases the sources of behavioral plasticity have been shown to be alterations in information transfer at specific points in the relevant neural pathways (Castellucci et al. 1970; Kupfermann et al. 1970; Zucker 1977).

Two forms of plasticity of the gill-withdrawal reflex of *Aplysia*— habituation and sensitization—are the result of changes in transmitter release from mechanoreceptor sensory neurons that are presynaptic to the interneurons and motor neurons of the reflex pathway (see Fig. 1). Habituation, a decline in reflex responsiveness that occurs when a stimulus is presented repeatedly, results from depression of transmitter release at the sensory neuron synapses (Castellucci and Kandel 1974). This form of plasticity is called homosynaptic, since it is restricted to the stimulated synaptic pathway and results from a change in the synapse itself. Sensitization, an increase in responsiveness that follows the presentation of a novel stimulus, is caused by presynaptic facilitation at the same synapses (Castellucci and Kandel 1976). There is evidence that the neurotransmitter that causes presynaptic facilitation is serotonin or a related substance, and that this

*Work presented here was done in collaboration with E. Kandel and supported by NIH grants NS12744 and GM23540. The help of L. Katz in preparing the figures is gratefully acknowledged.

neurotransmitter enhances transmitter release from sensory neurons by causing an increase in the concentration of cAMP in the terminals of these cells (Cedar et al. 1972; Cedar and Schwartz 1972; Brunelli et al. 1976). The experimental results presented here suggest that homosynaptic depression and presynaptic facilitation at these synapses are caused by modulation of the presynaptic I_{Ca} that is responsible for transmitter release.

HABITUATION: HOMOSYNAPTIC DEPRESSION AND I_{Ca} INACTIVATION

• Homosynaptic depression and presynaptic facilitation can be studied in a single module of the reflex pathway made up of one sensory

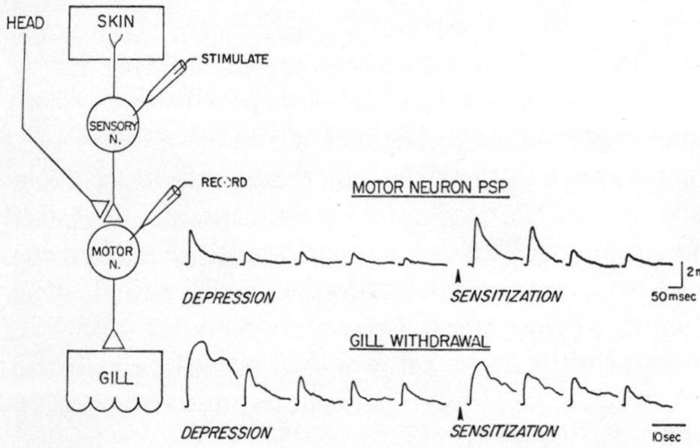

Figure 1
The gill-withdrawal reflex pathway and the behavioral and synaptic aspects of habituation and sensitization. Tactile stimulation of the siphon skin activates sensory neurons that connect to gill motor neurons (*left*). When the skin is stimulated at low frequency, the gill-withdrawal response decrements, or habituates, whereas head shock causes the response to increase, or sensitize (*right, lower traces*). In the reduced preparation, spikes are elicited by intracellular stimulation of sensory neurons to simulate tactile stimulation, intracellular recording of postsynaptic potentials replaces measurement of gill withdrawal, and electrical stimulation of a connective from the head replaces head shock (*left*). Postsynaptic potentials show both decrement with repeated sensory neuron stimulation and enhancement after nerve stimulation (*right, upper traces*), corresponding to the behavioral plasticity shown by the gill withdrawal itself. (Gill-withdrawal data from Pinsker et al. 1970; PSP data from Castellucci and Kandel 1976.)

neuron and a neuron postsynaptic to it (Fig. 1). Intracellular depolarization of a sensory neuron elicits action potentials that cause monosynaptic EPSPs in several interneurons and gill motor neurons. Repeated stimulation of the sensory neurons at 0.1 Hz results in homosynaptic depression of the EPSPs (Fig. 1; Castellucci and Kandel 1974). During homosynaptic depression in normal seawater medium, there is no change in the V_R or R_m of the presynaptic cell or in the configuration of its action potentials. However, when neurons are bathed in seawater containing TEA, which broadens action potentials by blocking some of the I_K that normally causes them to repolarize, depression of the EPSP is accompanied by progressive narrowing of presynaptic action potentials, and recovery of the EPSP after rest is accompanied by reversal of this narrowing (Fig. 2). Because in TEA the inward current during the later part of action potentials is carried by Ca^{++} (Horn and Miller 1977; Klein and Kandel 1978), and because Ca^{++} entry is essential for transmitter release (Katz and Miledi 1967), a reasonable hypothesis is that a decline in presynaptic Ca^{++} entry with successive action potentials contributes to the depression of transmitter release that underlies habituation. No change is observed in action potentials in normal seawater because the effect of I_{Ca} on V_m is obscured by the much larger I_{Na} and I_K. In the presence of TEA, enough I_K is blocked to allow I_{Ca} to play a significant role in determining the shape of the late part of the action potential, when a large part of the I_{Na} has been inactivated as well. Changes in I_{Ca} thus can be detected as changes in the duration of action potentials.

However, the abbreviation of I_{Ca} with repeated stimulation in the presence of TEA conceivably could be caused by the progressive increase of an opposing I_K. To examine I_{Ca} under more controlled conditions, sensory neurons were voltage-clamped in media in which both I_{Na} and I_K were eliminated as much as possible. In one set of experiments, intact neurons were depolarized briefly at 0.1 Hz in seawater containing TTX to block I_{Na} and also containing a high concentration of TEA to block I_K. EPSPs were recorded postsynaptically. In these experiments, the progressive decline of the EPSP correlated with a parallel decline in the presynaptic I_{Ca} (Klein et al. 1980). In these experiments the current flowing in the presynaptic terminals contributes little, if at all, to the total current recorded in the cell body. The parallel between cell-body currents and transmitter release nonetheless suggests that the currents in the cell body and in the terminals are affected in similar ways by repeated depolarization. The effect of voltage-clamp depolarization of the cell body on the terminals is discussed further below.

In other experiments, sensory neuron cell bodies were isolated from their processes to maximize voltage-clamp control and additional I_K-blocking agents were introduced. Repeated depolarization of the isolated sensory neurons resulted in a decrease in the inward current (see Fig. 4A, left) similar to that in the intact cell. The simplest interpretation of these findings is that inactivation of g_{Ca} occurs in response to repeated depolarizaton (see also Tillotson and Horn 1978). It is unclear as yet whether inactivation in the sensory neurons is caused by V_m changes per se, by entry of divalent cations (Tillotson 1979 and this volume), or by some other factor. The correlation between I_{Ca} inactivation and EPSP decrement suggests that progressive inactivation of I_{Ca} may contribute to the homosynaptic depression underlying habituation. To evaluate the quantitative contribution

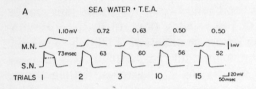

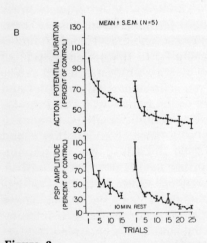

Figure 2

Decrease in Ca^{++} plateau potential duration during homosynaptic depression of the monosynaptic EPSP. Sensory neuron spikes and resulting EPSPs elicited at 0.1 Hz in high-divalent cation seawater containing 0.1 M TEA. (A) Sensory neuron action potentials narrow in parallel with depression of the EPSP during repeated stimulation. (B) Average action potential duration and EPSP amplitude (based on five preparations) during a first habituation run and during a second run after a 10-min rest. (Reprinted, with permission, from Klein et al. 1980.)

of I_{Ca} inactivation to homosynaptic depression, it will be necessary to examine the correlation between I_{Ca} and EPSP size in experiments in which both the duration and the frequency of presynaptic depolarizations are varied over a wider range.

SENSITIZATION: PRESYNAPTIC FACILITATION AND I_K MODULATION

• Electrical stimulation of one of the connectives, which carries sensitizing information from the head of *Aplysia*, causes presynaptic facilitation at the synapses of the sensory neurons in the gill-withdrawal reflex pathway (Castellucci and Kandel 1976). In addition, several changes occur in the membrane properties of the sensory neurons in normal seawater: the resting g_m decreases, the membrane often depolarizes slightly, and action potentials broaden by about 10%. In the presence of TEA, action potential broadening is much greater, often reaching several hundred percent. Broadening is correlated with facilitation of the monosynaptic EPSP recorded in follower cells (Klein and Kandel 1978; Klein et al. 1980). The effects of connective stimulation can last from several minutes to an hour. The changes in the resting membrane properties in normal seawater and the prolongation of action potentials in the presence of TEA can also be elicited by extracellular application of serotonin or phosphodiesterase inhibitors or by intracellular injection of cAMP (Klein and Kandel 1978). Since evidence has been presented that a serotonin-mediated increase in intraneuronal cAMP causes presynaptic facilitation at the synapses of these cells (Brunelli et al. 1976), the findings that connective stimulation, serotonin, and cAMP also cause changes in their membrane properties suggest that these changes in membrane properties might be the mechanism by which transmitter release is enhanced in presynaptic facilitation.

All the changes in membrane characteristics of the sensory neurons can be explained as effects of a synaptically mediated decrease in I_K. If this current were present at the V_R, its reduction would give rise to a depolarization accompanied by a decrease in g_m. This current could also contribute to repolarization of action potentials, and its diminution would therefore lead to action potential broadening. Broadening of action potentials could result in presynaptic facilitation, since prolongation of action potentials presumably would cause prolongation of the concomitant voltage-dependent Ca^{++} influx that causes transmitter release.

Voltage-clamp analysis of the decrease in g_m caused by connective stimulation indicates that the current that is turned off is voltage-sensitive, that the reversal potential of the synaptic current is affected by $[K^+]_e$, and that elimination of I_K blocks the effect of serotonin on g_m (Klein and Kandel 1980; Klein et al. 1980). These findings imply that the effect of connective stimulation is in fact a decrease in g_m.

To demonstrate that presynaptic facilitation is caused by a decrease in g_k, it is necessary also to show that prolongation of action potentials after connective stimulation is caused by a decrease in I_K and that the action potential broadening observed in normal seawater can cause a significant increase in transmitter release. The first of these questions was addressed by voltage-clamping sensory neurons at a V_h near the V_R and introducing brief depolarizing commands up to the approximate level reached by the peak of an action potential. At the same time, monosynaptic EPSPs caused by the presynaptic depolarizations were recorded in follower cells. In these experiments, voltage-clamp control is satisfactory only in the cell body and its immediate vicinity. The presynaptic terminals are not well-controlled, and EPSPs are caused by action potentials in the terminals. Thus, currents observed in the cell body are not causally related in a simple way to transmitter release from the terminals but they can reflect, indirectly, changes in the terminal currents.

In solutions containing no drugs, repeated depolarization at 0.1 Hz is accompanied by depression of the EPSP, but there is no change in the presynaptic currents (Fig. 3, A [left] and B). No change in the presynaptic currents is to be expected if, as discussed above, homosynaptic depression is associated with inactivation of a relatively small I_{Ca}. After connective stimulation, however, an increase in the monosynaptic EPSP is accompanied by a decrease in the peak outward current of the presynaptic neuron (Fig. 3, A [right] and B). A similar decrease in net outward current occurs after addition of serotonin. The decrease in net outward current after nerve stimulation is responsible for the broadening of action potentials in normal seawater (Klein and Kandel 1978). This change in the net current could reflect a change in any of the ionic currents activated by the depolarization command. Further analysis of the effects of connective stimulation and serotonin was necessary to specify which current (or currents) is affected.

Earlier experiments in which action potentials were examined in the presence of TEA indicated that connective stimulation and serotonin incubation do not affect I_{Na} (Klein and Kandel 1978). The remaining candidates for the locus of action of synaptic stimulation and serotonin are therefore I_{Ca} and I_K. To distinguish between these two possibilities, the effect of serotonin was tested in the presence and in the absence of I_K. In one set of experiments, sensory neuron cell bodies were separated from their processes to maximize voltage-clamp control, and TEA^+ was substituted for Na^+, Ba^{++} for Ca^{++}, and 4-AP added to the bathing solution. These pharmacological treat-

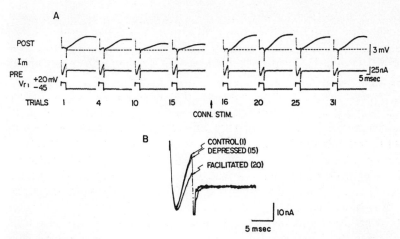

Figure 3
Transient currents during homosynaptic depression and presynaptic
facilitation. (A) Sensory neuron voltage clamped in high divalent
cation seawater and EPSP recorded postsynaptically. Presynaptic V_m
stepped from -45 mV to $+20$ mV at 0.1 Hz (*lower traces*), eliciting
transient inward and outward currents in the sensory neuron (*middle traces*) and EPSPs in the follower cell (*upper traces*). EPSPs
undergo homosynaptic depression in trials 1–15. Connective stimulation produces facilitation of the EPSP (*top*) and reduction in presynaptic outward current (*middle*) in trials 16–31. (B) Three traces
from A superimposed to illustrate lack of significant change in
currents during synaptic depression (traces 1 and 15) and reduction
of outward current associated with presynaptic facilitation (trace 20).
(Reprinted, with permission, from Klein and Kandel 1980.)

ments have been shown to reduce I_K (Thompson 1977). In some
experiments, Ca^{++} was used instead of Ba^{++}. In all cases, inward
current carried by either Ca^{++} or Ba^{++} through the Ca^{++} channels was
not affected by serotonin (Fig. 4A). When these solutions were replaced with normal seawater, however, serotonin once again reduced
the net outward current presumably carried by K^+ (Fig. 4B).

   The combination of drugs used to block I_K might have some
additional effect, perhaps on the serotonin receptor or the Ca^{++} channel itself. In that case, it would be possible to interpret the results of
the above experiments in more than one way, and it would not be
possible to conclude with any certainty that serotonin acts exclusively on I_K. Elimination of I_K was therefore accomplished in a different
way, by removing K^+ from the interior of the cell. Intracellular K^+ was
replaced with impermeant Cs^+ using the antibiotic nystatin to effect
ion replacement (Russell et al. 1977; Tillotson and Horn 1978). In the
absence of I_K in these experiments, serotonin again had no effect on
the membrane currents of the sensory neurons. When K^+ was reintro-

duced into the cells by means of a KCl micropipette, serotonin once more was able to reduce the outward current (Klein and Kandel 1980; Klein et al. 1980). The above results imply that serotonin and, presumably, nerve stimulation act to reduce I_K and have no direct effect on I_{Ca}.

The next step in demonstrating that a reduction in I_K causes presynaptic facilitation involves showing that the prolongation of action potentials that is caused by the synaptically mediated reduction of I_K can account for the increase in EPSP amplitude that typically follows connective stimulation. Therefore, the relation between duration of presynaptic depolarization and EPSP amplitude was examined using voltage-clamp steps of different durations in the presence of I_K-blocking agents. Although only a small number of experiments were done and lack of voltage-clamp control of the terminals complicates this analysis, the results are consistent in showing that longer depolarizations produce larger EPSPs and that the dependence of EPSP amplitude on presynaptic pulse duration is steeper for briefer pulses (Klein et al. 1980). Extrapolation of these results to action potentials in normal seawater suggests that the action potential prolongation that is usually observed after connective stimulation could account for the concomitant increase in transmitter release during presynaptic facilitation.

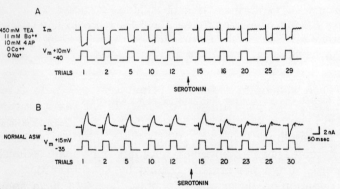

Figure 4
I_m in the presence and absence of I_K blockers: effects of repeated depolarization and serotonin application. (A) With I_K blocked, step depolarizations at 0.1 Hz (*bottom traces*) cause gradual decline in inward current (*top traces*). Addition of serotonin has no effect. (B) Protocol was repeated in the same cell after replacement of drug solution with normal seawater. Serotonin causes a decrease in the outward currents elicited by depolarizing steps and an inward shift in the holding current. Sensory neuron soma isolated by ligation. (Modified from Klein and Kandel 1980.)

EPSP CONFIGURATION IN PRESYNAPTIC FACILITATION AND HOMOSYNAPTIC DEPRESSION

• If presynaptic facilitation is caused by prolongation of presynaptic action potentials, then facilitated EPSPs might be expected to have longer rising phases than control EPSPs. When presynaptic pulse duration is increased under voltage clamp, the resulting enhanced EPSPs do, in fact, have longer rise times than the smaller EPSPs evoked with shorter pulses (Klein et al. 1980). In contrast, increasing the EPSP amplitude by increasing the $[Ca^{++}]_e$ does not result in a change in EPSP shape (Castellucci and Kandel 1974). Thus, if EPSP rise times were significantly longer after connective stimulation, this would be suggestive evidence that a prolongation in terminal action

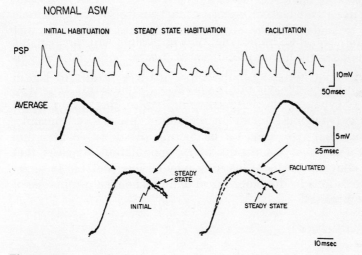

Figure 5
Comparison of EPSP shape during habituation and sensitization. Monosynaptic EPSPs were evoked with sensory neuron action potentials fired once every 10 sec in normal seawater. Blocks of five EPSPs at the beginning and end of the habituation run and after connective stimulation (*uppermost traces*) were averaged using a signal averager (*middle traces*). The averages were scaled so that their peak amplitudes were the same and were then superimposed in pairs (*bottom traces*). The traces at the bottom left show superimposition of the averages taken from the beginning and end of the habituation run. There was a considerable decrement in EPSPs during habituation, but their shape remained constant. The traces on the bottom right show superimposition of the averages of the habituated EPSPs (*solid trace*) and the facilitated EPSPs (*dashed trace*). The facilitated EPSPs have a longer time-to-peak and a slower initial decline than the habituated ones. (Reprinted, with permission, from Klein et al. 1980.)

potentials occurs and that this prolongation might contribute to the increase in EPSP amplitude. EPSPs elicited by firing action potentials in sensory neurons in normal seawater do exhibit longer times-to-peak after presynaptic facilitation but show no change in shape during homosynaptic depression (Fig. 5). This finding confirms a prediction based on the hypothesis that presynaptic facilitation is caused by action potential broadening in the sensory neurons. The fact that EPSPs do not change in shape during homosynaptic depression is consistent with the hypothesis that depression is caused by I_{Ca} inactivation, rather than by a change in configuration of the presynaptic action potential.

CONCLUSION

• Despite the fact that homosynaptic depression and presynaptic facilitation apparently have different mechanisms, both forms of synaptic plasticity are associated with changes in the voltage-dependent Ca^{++} influx into the presynaptic neurons. I_{Ca} modulation underlies other forms of synaptic plasticity as well (Shapiro, this volume), and it may be a general mechanism for regulation of synaptic transmission. The fact that two forms of behavioral plasticity—habituation and sensitization—may be results of I_{Ca} modulation suggests that other, more complex forms of learning may be interpretable in terms of similar alterations in the biophysical and biochemical properties of neuronal membranes.

REFERENCES

Brunelli, M., V. Castellucci, and E. R. Kandel. 1976. Synaptic facilitation and behavioral sensitization in *Aplysia*: Possible role of serotonin and cyclic AMP. *Science* **194**: 1178.

Castellucci, V. F. and E. R. Kandel. 1974. A quantal analysis of the synaptic depression underlying habituation of the gill-withdrawal reflex in *Aplysia*. *Proc. Natl. Acad. Sci.* **71**: 5004.

―――――. 1976. Presynaptic facilitation as a mechanism for behavioral sensitization in *Aplysia*. *Science* **194**: 1176.

Castellucci, V., H. Pinsker, I. Kupfermann, and E. R. Kandel. 1970. Neuronal mechanisms of habituation and dishabituation of the gill-withdrawal reflex in *Aplysia*. *Science* **167**: 1745.

Cedar, H. and J. H. Schwartz. 1972. Cyclic adenosine monophosphate in the nervous system of *Aplysia californica*. II. Effect of serotonin and dopamine. *J. Gen. Physiol.* **60**: 570.

Cedar, H., E. R. Kandel, and J. H. Schwartz. 1972. Cyclic adenosine monophosphate in the nervous system of *Aplysia californica*. I. Increased synthesis in response to synaptic stimulation. *J. Gen. Physiol.* **60**: 558.

Horn, R. and J. J. Miller. 1977. A prolonged, voltage-dependent calcium permeability revealed by tetraethylammonium in the soma and axon of *Aplysia* giant neuron. *J. Neurobiol.* **8:** 399.

Katz, B. and R. Miledi. 1967. The timing of calcium action during neuromuscular transmission. *J. Physiol. (Lond.)* **189:** 535.

Klein, M. and E. R. Kandel. 1978. Presynaptic modulation of voltage-dependent Ca^{2+} current: Mechanism for behavioral sensitization in *Aplysia californica. Proc. Natl. Acad. Sci.* **75:** 3512.

————. 1980. Mechanism of calcium current modulation underlying presynaptic facilitation and behavioral sensitization in *Aplysia. Proc. Natl. Acad. Sci.* (in press).

Klein M., E. Shapiro, and E. R. Kandel. 1980. Synaptic plasticity and the modulation of the Ca^{2+} current. *J. Exp. Biol.* (in press).

Kupfermann, I., V. Castellucci, H. Pinsker, and E. R. Kandel. 1970. Neuronal correlates of habituation and dishabituation of the gill-withdrawal reflex in *Aplysia. Science* **167:** 1743.

Pinsker, H., I. Kupfermann, V. Castellucci, and E. R. Kandel. 1970. Habituation and dishabituation of the gill-withdrawal reflex in *Aplysia. Science* **167:** 1740.

Russell, J. M., D. C. Eaton, and M. S. Brodwick. 1977. Effects of nystatin on membrane conductance and internal ion activities in *Aplysia* neurons. *J. Membr. Biol.* **37:** 137.

Thompson, S. H. 1977. Three pharmacologically distinct potassium channels in molluscan neurones. *J. Physiol. (Lond.)* **265:** 465.

Tillotson, D. 1979. Inactivation of Ca^{++} conductance dependent on entry of Ca^{++} ions in molluscan neurons. *Proc. Natl. Acad. Sci.* **76:** 1497.

Tillotson, D. and R. Horn. 1978. Inactivation without facilitation of calcium conductance in caesium-loaded neurones of *Aplysia. Nature* **273:** 312.

Zucker, R. S. 1977. Synaptic plasticity at crayfish neuromuscular junctions. In *Identified neurons and behavior of arthropods* (ed. G. Hoyle), p. 49. Plenum Press, New York.

Key to Terms*

activation curve, 21
aequorin, 30
arsenazo, 30
capacity transient, 20
channels, 11
chemically gated channel, 12
command voltage, 17
constant-field assumption, 24
constant-field equation, 24
corner frequency, 28
cutoff frequency, 28
dialysis, 26
differential pair, 30
displacement current, 19
dose-response curve, 21
error signal, 17
Eyring rate theory, 24
fluctuation analysis, 27
flux equation, 24
gating, 12, 15
Goldman-Hodgkin-Katz (GHK) equation, 24
instantaneous I-V curve, 23
internal perfusion, 26
ion-hopping models, 24
ion-sensitive electrodes, 30
leakage resistance, 20

leakage subtraction, 20
leak correction, 20
Lorentzian, 28
membrane conductance, 16
membrane protein metabolism, 15
membrane protein regulation, 15
modulation, 15
neurotransmitter, 12
noise, 27
patch recording, 29
permeabilities, 24
permeation, 12
point clamp, 18
rectification, 23
relaxation time constant, 21
selectivity, 12
separation of currents, 22
series resistance, 17
series resistance artifacts, 18
setting time constant, 19
single-channel recording, 29
single-electrode voltage clamp, 18
single-time-constant spectrum, 28
space clamp, 18
spectrum, 28
steady-state I-V curve, 21
surface-charge effects, 25

*The terms listed here are defined and discussed in detail on the pages indicated.

223

Index